# A NOTE ON THE AUTHOR

Natalie Starkey is a geologist, cosmochemist and science communicator. Natalie's doctorate at Edinburgh University on the geochemistry of Arctic volcanoes saw her travelling to the volcanic lava-fields of Iceland and the ancient volcanoes of northern Scotland, and she also spent time as a volcanologist on the island of Montserrat in the Caribbean. Later, Natalie's postdoctoral research expanded to include the analysis of rock samples from space, which led to her first popular science book, *Catching Stardust* (Bloomsbury Sigma, 2018).

Natalie received a British Science Association Media Fellowship in 2013 to work with the *Guardian*. She has been a science host on Neil deGrasse Tyson's popular StarTalk Radio, and is now Science Media Producer for Chemistry World at The Royal Society of Chemistry.

@starkeystardust / nataliestarkey.com

## Some other titles in the Bloomsbury Sigma series:

# FIRE

## AND

# ICE

## VOLCANOES OF THE
## SOLAR SYSTEM

Natalie Starkey

BLOOMSBURY SIGMA
LONDON • OXFORD • NEW YORK • NEW DELHI • SYDNEY

BLOOMSBURY SIGMA
Bloomsbury Publishing Plc
50 Bedford Square, London, WC1B 3DP, UK
29 Earlsfort Terrace, Dublin 2, Ireland

BLOOMSBURY, BLOOMSBURY SIGMA and the Bloomsbury Sigma logo are
trademarks of Bloomsbury Publishing Plc

First published in the United Kingdom in 2021.
This edition published 2023.

Photo credits (t = top, b = bottom, l = left, r = right, c = centre)

Colour section: P. 1: © NASA / JPL. P. 2: © Natalie Starkey (t, c, b). P. 3: © NASA
/ JSC (t); © US Geological Survey (b). P. 4: © NASA / Johns Hopkins University
Applied Physics Laboratory / Carnegie Institution of Washington (t); © NASA / JPL
(c); © NASA / JPL / Universities Space Research Association / Lunar & Planetary
Institute (b). P. 5: © NASA / JPL-Caltech / University of Arizona / ASU
(t and inset); © NASA / JPL / Space Science Institute (c); © NASA / JPL (b).
P. 6: © NASA / JPL-Caltech / University of Nantes / University of Arizona (t); ©
NASA / JPL-Caltech / US Geological Survey (c); © NASA / JPL (b). P. 7: © NASA
/ Johns Hopkins University Applied Physics Laboratory / Southwest Research
Institute (tl and tr); © NASA / JPL-Caltech / UCLA / MPS / DLR /
IDA (c and b). P. 8: © David Blanchflower (t); © ESA / Space-X, Space
Exploration Institute, CC BY-SA 3.0 IGO (c); © David Blanchflower (b).

A catalogue record for this book is available from the British Library

Library of Congress Cataloguing-in-Publication data has been applied for

ISBN: PB: 978-1-4729-6040-5; eBook: 978-1-4729-6038-2

2 4 6 8 10 9 7 5 3 1

Typeset by Deanta Global Publishing Services, Chennai, India
Printed and bound in Great Britain by CPI Group (UK) Ltd, Croydon CR0 4YY

Bloomsbury Sigma, Book Sixty-Nine

To find out more about our authors and books visit www.bloomsbury.com
and sign up for our newsletters

For C.W., K.W. and Charlie

# Contents

# Preface

Like many of you, when I first studied the planets, it was accepted that there were nine of them. Pluto was still classed as a fully fledged planet, yet we only had a very blurred image of its surface and knew even less about what went on there. I was school age at the time the world started learning in more detail about the planets in the outer Solar System, including their many and varied moons. This was because we were in the process of visiting them with spacecraft for the first time and analysing the data being returned. What was found was completely unexpected. None of the places that our spacecraft visited and observed looked and behaved as we'd imagined. These places were worlds in their own right, with active surfaces being the norm, and the concept of completely barren, lifeless planets became more and more unlikely.

I've always been fascinated by the planetary objects that orbit closest to the Sun, the so-called terrestrial, or rocky, planets: Mercury, Venus and Mars, and our only natural satellite, the Moon. In terms of distance they are close to Earth, but these objects are also close to our planet in many other ways, being made of the same rocky starting materials and hosting similar-looking features on their surfaces, such as mountains and valleys. I was always amazed that these worlds once hosted volcanoes that were just like some of those we have on Earth today; they had erupted hot, molten rock that reformed their surfaces. This is the case even for the Moon, which today is an apparently dead or dormant grey rock. Yet, while we've explored the Moon in some detail with spacecraft, including landers and orbiters, and we've even sent humans there, the idea of humans investigating the other rocky planets is still some time in the future, if it ever happens. These objects might be close to us, but the conditions that exist on their surfaces, whether extremes of pressure or temperature, make the prospect of exploring them very challenging.

This brings me to the objects beyond the inner Solar System, the so-called gas and ice giants of the outer Solar System: Jupiter, Saturn, Uranus and Neptune, and the many moons that surround them. These enormous planets are nothing like our own; they don't have solid surfaces for us to land upon even if we managed to design a craft that could withstand the extreme conditions of such worlds. Some of the most important missions for improving our understanding of these places were the un-crewed Voyager spacecraft that ventured out as part of NASA's so-called 'Grand Tour of the Solar System'. Voyager 2 launched just a few weeks ahead of Voyager 1 in 1977, and because they were sent on different trajectories, Voyager 1 reached its target first. They explored Jupiter and Saturn initially, before the missions were extended to take in Uranus and Neptune as well. Today, they continue their journey into the unknown, exploring the region beyond our planets and at the outer limits of the Sun's sphere of influence. They are currently the most distant human-made objects from Earth.

The two Voyager spacecraft beamed back images of planets and moons of which we knew very little, such were the huge distances of these objects from our own planet. To date, they also remain the only spacecraft to have visited Uranus and Neptune. Despite the fleeting fly-bys provided by the Voyager spacecraft, their visits gave us a vast insight into what these places look like and how they behave, with some surprising revelations. As Voyager swung by the largest planetary bodies in our Solar System, finding out, for example, that Jupiter's Great Red Spot was a complex raging storm, we found that our planet was not the only one with an interesting, and long, story to tell.

Voyager discovered new moons that we hadn't even known existed; it recorded the temperatures of the planets; it made measurements of them, including the phenomenally thin rings of Saturn. But most exciting of all was the discovery that our planet is not alone in housing active volcanoes. The Voyager craft observed eruptions occurring on Io, one of Jupiter's moons. In fact, between the two Voyager spacecraft, nine eruptions were seen, and it is thought that others occurred in between the two fly-bys.

Io is a planetary body seemingly so unlike our own. It sits next to Jupiter and its daily existence is seriously influenced by this behemoth of the Solar System. Yet, despite this, Io shows similarities to our own planet. As far as we know, it is the only body in the outer Solar System to erupt molten rock.

Of course, scientists already knew about the existence of many of these moons, as they could see them with telescopes from Earth. As early as 1610, Galileo Galilei saw a pinpoint of light in the sky, and discovered the first moon orbiting another planet. This was Io, orbiting Jupiter, which was originally named Jupiter 1. Its discovery, and that of the other Galilean moons, led to the understanding that planets in the Solar System orbit the Sun, rather than everything revolving around Earth. However, because of their small size and great distance from us, relatively little was known about their surfaces and, in particular, the fact that they were active. The exciting thing is, in many ways, these moons are just as interesting as the large planets that they orbit, revealing other pieces of the multifaceted and complicated Solar System story.

Although we were learning that Earth is not the only place in the Solar System to host hot, active volcanoes, we certainly weren't expecting to find volcanoes on the surface of the icy worlds. Yet, along with all the other phenomenal images that were beamed back by Voyager, there were some scientific gems of information that remained hidden for a few more decades. One of these was the fact that Enceladus, the sixth-largest moon of Saturn and a distinctly icy place, was volcanically active, but not in the sense that we would consider 'normal'. What Voyager did reveal straight away was a 'fresh' surface, one that was apparently geologically active. Enceladus' surface was scarred with very few impact craters, instead displaying faults and valleys. These landscapes hinted to scientists that this icy world had an interesting story to tell, one that looked as if Enceladus had been geologically active in the relatively recent past.

However, proof of this assumption didn't come for another 30 years, in 2005, when Enceladus was revisited by NASA's Cassini spacecraft, which had set out to explore the environs of Saturn and its many moons. Cassini discovered so much about the Saturnian system, far too much to be covered here. But something I do want to mention that is of most relevance to the topic of this book, is the icy plume activity of Enceladus. Cassini captured images of geyser-like jets of icy particles being pumped from Enceladus' south polar region. This was proof that this little moon is active today, releasing substances from its interior out into space. Once this was discovered, the images that Voyager had captured decades earlier were re-examined and re-processed, revealing that Enceladus' plumes were also active then, they just hadn't made themselves so obvious in the images and data returned at the time (you had to be looking for them to know they were there). So Enceladus' geyser activity was not a sporadic, fleeting feature, but one that seemed to have been active over a much longer period.

The small size of Enceladus had scientists completely fooled. What they expected was a dead world, one that had long been frozen solid, but while Enceladus' surface is indeed covered by ice, the presence of icy plumes emanating from within it strongly hinted at the presence of a liquid ocean beneath its cold, solid carapace. This meant that Enceladus, despite its extreme distance from the Sun, retained some internal heat; enough in fact to heat a subsurface ocean. Once again, scientists were learning to expect the unexpected. We'll focus on Enceladus in a few other chapters in this book because it is an important icy world that has a lot to teach us about volcanic activity beyond our own planet.

For now, we should consider the implications of this particular mix of features:

- the presence of liquid water;
- a source of heat;
- the resultant movements of material from the inside to the outside of the moon.

The combination of these factors means that Enceladus has the potential to host life. Yet life at this great distance from the Sun was traditionally thought to be impossible, the location being so far removed from the so-called 'habitable zone' where Venus, Earth and Mars are found, which is not too close and not too far from the Sun. These new findings on Enceladus awakened the possibility that we weren't alone in the Solar System. Those little green aliens on Mars that I'd heard about when I was a child were going to need to move to a new home a bit further afield, possibly around Jupiter or Saturn.

In 2015, nearly 40 years after the Voyager spacecraft revealed the true beauty of the large planets and their satellites that share the space around our Sun, the New Horizons mission arrived at Pluto. It had never been visited by Voyager or any other spacecraft. Pluto sits within the Kuiper Belt, a region of our Solar System beyond the orbits of the planets that hosts millions of small icy objects. Pluto is often referred to as the 'King of the Kuiper Belt' because of its larger size in comparison to the objects surrounding it, but in reality it should share this title with a number of other large objects that have been discovered in the region in more recent years. Nevertheless, what was revealed at this most far-flung of former planets was another active world, which was potentially even volcanic, yet it was a space object that had long been thought of as not much more than a dead lump of ice. To this day, Pluto provides a conundrum for scientists. How is activity possible this far from the Sun? Where do icy worlds such as Pluto manage to find their heat? There is still much to learn about Pluto and the many other icy worlds that surround it. Discovering more about these places will be best achieved with future missions that can spend an extended time at such locations, using orbiters and landers to explore their surfaces in detail. These missions will build on the fleeting, yet wholly revealing fly-bys of craft such as Voyager and New Horizons, which paved the way and revealed the true nature of our active and evolving Solar System.

Finding volcanic activity, albeit from a volcano composed of ice, on worlds that were thought to be frigid and long-dead forced scientists to reconsider volcano classification. These icy outer Solar System worlds revealed that not all volcanic activity necessarily resembled that on Earth. In fact, considering the number of active, icy planetary bodies that are now known to exist, it might be that Earth is the oddball of the Solar System, biasing our view of what a volcano should look like!

While Mars, the Moon and Venus have volcanoes that show striking similarities to some of our own Earthly varieties, when other spacecraft captured images of activity on cold, icy planets, it took a leap of faith to even call this activity volcanic. In fact, it wasn't obvious at first what the scientists were seeing. The plumes of activity captured at Enceladus seemed to resemble geysers on Earth, where hot spring water spouts from the ground, and did not look like something we technically class as a volcano, despite it being a 'volcano-related' feature.

But does it matter? Surely a volcano is a volcano whether it is on Earth or on another planet or moon in deep space? Volcanoes are a part of the efforts a planetary body makes to cool itself down, releasing excess heat into space. However, one of the issues here is that the inner workings of other planets and moons – the parts that drive their eruptions and activity – are different from Earth. Each planetary body is composed of slightly different concoctions of Solar System materials. Some are not even rocky, instead being made of highly compressed ices that are just as hard as solid rock at that body's temperature. Not to mention that the planetary bodies that make up the Solar System all sit at different distances from the Sun, existing in vastly different environments. So, while scientists can apply some of their knowledge about Earth's volcanoes to those in space, the conditions on other planetary bodies are just too different for simple comparisons to be made.

Trying to forecast how an eruption on another planetary body might proceed when we don't have a clear idea of what it is made of, or what is its internal structure, is nigh on impossible, even if we have a good idea about the temperatures and pressures existing at its surface.

Just learning that some Solar System volcanoes exist at all has been a surprise. If we are to really understand what makes these extraterrestrial volcanoes tick, we are going to need future space missions to visit and investigate them directly. Nevertheless, we can gain an appreciation of alien worlds just by using images and collecting other basic scientific information as spacecraft pass by. The simplest next step is to compare the features we've observed to those on Earth. This is known as comparative planetology and is a powerful arm of scientific study, as it allows us to make inferences about worlds we've barely visited. These are first results that can then be tested once we can go there and take a close-up look. We'll look at this in more detail in future chapters.

Here on Earth we are fortunate that humans can explore volcanoes in person, trampling over their smouldering surfaces, poking them, and collecting samples of their gases and rocks. In space, we are not afforded these luxuries, so scientists – once they've learned of the potential existence of a cosmic volcano – must instead rely on measurements and images of 'field sites' taken remotely, often by a spacecraft many hundreds or thousands of kilometres away.

Nevertheless, being unable to send humans to investigate space's volcanoes is not all bad. After all, volcanology is a dangerous profession. Recently, in an era of improved health and safety regulations and a marginally better understanding of when a volcano is likely to erupt, volcanologists have become a slightly more protected species. Despite this, there have been many tragic deaths as a result of volcanic activity. Most deaths associated with volcanoes occur within 1 kilometre (0.6 miles) of the volcanic peak, highlighting the danger in working as a field scientist where sampling is often required near, or at, the summit of an active volcano.

Of course, it's not always volcanologists who are in danger *per se*; sometimes it is experienced volcano guides, tourists or simply residents. These people might have enthusiastically clambered too close to a volcano, unexpectedly being overcome with gaseous volcanic exhaust, or becoming trapped in a valley not knowing a scorching hot ash flow was on its way. Interestingly, the majority of deaths in this category come during a volcano's quiescent times, when it is perhaps thought to be safe to visit. This reinforces the fact that volcanoes are unpredictable and present some seemingly invisible threats. A poignant example is the death of six tourists in five separate incidents at Asosan volcano in Japan over a nine-year period from 1989 to 1997. During this time, the volcano was in a quiescent phase, undergoing a relatively normal – and what was thought to be 'safe' – amount of degassing (a term describing the natural release of magmatic gases from underneath the volcano). However, even small amounts of degassing can still be hazardous to certain people, and it was later found that the tourists who died were especially susceptible to changes in air quality due to pre-existing pulmonary conditions. These tragic incidents show just how important it is to heed the warnings about visiting active volcanoes, even when they are in a seemingly inactive phase.

One of the more recent times this was evident was in the eruption of the Whakaari volcano on White Island in New Zealand. Around 10,000 people visit this volcano every year. It is New Zealand's most active, which has been built up over the last 150,000 years from repeated eruptions in near-continuous volcanic activity. Towards the end of 2019, scientists as part of GeoNet, New Zealand's seismic monitoring agency, noticed that the volcano was experiencing increased activity. As a result, the alert level was raised from 1 to 2, indicating that a moderate eruption was due (on a scale from 0 to 5 where 5 is a major eruption). The eruption that came at the start of December 2019 was not particularly large, described by one scientist as a 'throat clearing'. Nevertheless, there were 47 people visiting the island at the time, most of them tourists on a volcano tour. Thirteen members of this group were killed instantly, with the others requiring hospitalisation for severe burns, some of which proved fatal later on. Two people were never found and presumed dead. We might question what went wrong here? The tour company, White Island Tours, published a statement on their website before the incident that read:

> Whakaari/White Island is currently on Alert Level 2. This level indicates moderate to heightened volcanic unrest, there is the potential for eruption hazards to occur. White Island Tours operates through the varying alert levels, but passengers should be aware that there is always a risk of eruptive activity regardless of the alert level. White Island Tours follows a comprehensive safety plan which determines our activities on the island at the various levels.

The information provided by the tour company was accurate and up to date, but we have to wonder whether the tourists that travelled under these conditions really knew what they were getting themselves into and just how unpredictable and dangerous volcanoes like Whakaari can be. All we know is that this won't be the last such incident. These fiery mountains, however fascinating and beautiful, are beasts that cannot be tamed.

The risk that others face in the study of volcanoes is, morbidly perhaps, sometimes also an inspiration. You may wonder why anyone bothers to study these dangerous, fiery mountains. Perhaps part of the excitement for a volcanologist is seeing the birth of new rocks: molten rock pouring out as a lava flow, then cooling to become the youngest piece of Earth. In collecting and studying these rocks, we gain information about the Earth's interior and how our planet became the place it is, but we can also use these rocks to forecast what the Earth will do in the future, including whether that particular volcano will erupt again soon.

When I was at school, I recall learning about the deaths of Katia and Maurice Krafft, extremely experienced volcano documenters who spent their lives travelling the globe chasing volcanic eruptions. They regularly risked their lives – although they may not have seen it this way – to film and photograph active volcanoes. They weren't just interested in the volcanoes themselves. In particular, the Kraffts focused on documenting the potentially deadly flows emanating from volcanoes. It's certain that many modern-day volcanologists have been inspired by Katia and Maurice and have learnt a great deal about the mechanisms of eruptions from their iconic images. Unfortunately, the Kraffts were killed on Mount Unzen in Japan in 1991, along with 41 other people, one of whom was a famous and highly experienced volcanologist called Harry Glicken. The large group were using a ridgeline near Unzen as a filming location, above a valley where pyroclastic flows – incandescent clouds of rock dust travelling at hundreds of kilometres per hour – had previously hurtled down. Their high location was deemed safe as the dense, hot flows had always clung to the valley, allowing the team to look down on them. However, in this incident, an especially large flow unexpectedly came down the valley, slightly less dense than the previous ones, which allowed it to sweep up over the ridge, inundating all who stood there with a searing cloud of rock dust and particles. The tragic loss of these experts was a stark reminder to the volcanology community that volcanoes are unpredictable and that even the most seasoned scientists, who have spent a lifetime studying them, are often in danger. Nevertheless, without these courageous and pioneering scientists, our knowledge of volcanoes would be so much poorer.

I don't consider myself to be particularly brave, but I too have traipsed on the side of active volcanoes in places such as the island of Hawai'i, Iceland and the Caribbean. I've even been close enough to see bubbling lava and hear the roar of a volcano as gases forcefully escape through its volcanic peak. I simply find them too fascinating to ignore, despite knowing the dangers of getting too close at the wrong time.

Volcanoes in space, of course, have yet to receive a visit from me, or anyone else for that matter. However, thanks to the many un-crewed space missions that have launched over the years, and continue to do so, we Earth-bound folk are privileged to see some of these distant volcanic worlds without having to leave the comfort and safety of our homes or laboratories. Clearly, volcanology is a dangerous profession, but so too is space exploration, and combining an astronaut and a volcanologist might result in a rather treacherous mix. Therefore, studying space volcanoes using robotic, un-crewed missions seems like a sensible idea, at least until we understand a little more about the icy or fiery monsters we are dealing with. There is still a lot to learn about the volcanoes that have been found in space so far, and there are probably many more to discover; it's a job best left to robotic spacecraft that can scan a potentially active planetary surface from a safe distance.

Yet it isn't only volcanoes in space that are hard to find. Of the thousands of volcanoes on Earth, we've only explored relatively few. One of the problems is not that they are too high to scale, or too dangerous to approach, but that so many are hidden in the cripplingly cold and pressurised depths of Earth's oceans. As a result, these volcanoes are all but inaccessible to humans. In recent decades some of these submarine volcanoes have been investigated by deep-sea robotic submersibles but, even so, relatively little is known about them in comparison to their counterparts that exist above sea level. Learning more about these volcanoes requires further technical investment and time. This is not dissimilar to the issues that humans face in exploring space, and it's part of the reason why we know so little about the Solar System's other volcanoes, the ones that aren't located on our special blue planet.

While we often focus on the exciting yet enormously destructive force of volcanoes, it is paying a disservice to them if we only concentrate on this negative aspect of their existence. Of course, volcanoes get their name on the map and become famous *because* of their eruptions, but it is these same events that are responsible for a whole variety of creative consequences too. Volcanoes literally build new land; they produce mountains and islands where once there was nothing. They can generate energy. They can even add a natural fertiliser to the surrounding land, often making their slopes particularly productive for farming. The latter is definitely a factor that increases the threat volcanoes may pose in the future, because people are keen to cultivate the newly fertile land for food production while, at the same time, putting themselves at risk.

Most importantly, it might be that life itself cannot exist without volcanoes. This is a contentious issue and scientists are still not sure of the exact mechanism that led to the start of life on Earth, where it came from or even exactly where it first took hold. Volcanoes certainly represent reasonable candidates for releasing the necessary ingredients to make a planet habitable, in particular the gases and volatiles such as water. Volcanic activity itself is a key indicator that a planet is 'alive', as volcanoes are a direct result of processes occurring within the deep interior of a planet. They are a manifestation of a planet attempting to cool itself from the inside out, by literally letting off steam. When we want to search for signs of life in space, the first places we should look at are those planets and moons that show, or have shown, volcanic activity, and from these we have so very much to learn.

# CHAPTER ONE

# Introduction

E arth, despite being known as the blue planet, is red-hot. We might not be aware of it but the centre of our planet, nearly 6,500 kilometres (4,000 miles) below the surface, is hotter than the surface of the Sun. Luckily for us, going about our daily lives atop the rigid, rocky layer that forms the Earth's crust – much like the skin of an apple – we are happily oblivious to the extreme temperatures deep below. However, it is thanks to the heat within our planet that Earth stays perfectly warm enough, yet provides the pleasant conditions necessary for life. Earth can be thought of as a giant hot-water bottle that is continuously – and magically – topped up with freshly warmed water. Its outer crustal shell controls the heat that works its way out from the interior to escape into space as the planet regulates its temperature. This flow, or movement, of heat from the deep Earth outwards is what drives activity on the surface, such as volcanic eruptions and earthquakes. In geological terms, these are broadly known as tectonic events: large-scale physical processes that affect the outer portion of the Earth. They are important to us because they are responsible for the building, and destruction, of sections of the Earth's outer shell, which is the part we live on. But Earth is not alone in this respect. Tectonic processes have also occurred on other planets and moons within our Solar System and they too have, or have had, warm interiors in need of cooling, resulting in activity at their surfaces.

## What is a volcano?

It would probably be useful to define the term 'volcano'. This may seem a bit unnecessary as most people these days probably think they have a reasonable understanding of what a volcano is, on Earth at least. Usually we think of a volcano as simply an opening or fracture in the Earth's crust where hot, liquid rock, or magma, bubbles up. The dictionary describes a volcano as a conical mountain or hill, with a crater or vent through which lava or gas escapes. This definition certainly seems to work for many volcanoes that instantly spring to mind, even if you might also imagine them with snowy peaks, like the classic images of fiery mountains such as Mount Fuji in Japan. However, when we move out into the Solar System, the definition describes only a handful. A problem with volcano science is that it is skewed towards Earth, because for many years that is the only place we knew volcanoes existed. Even when we started to learn about volcanoes on other worlds, we still had rather limited information about them, having not physically been to their slopes to obtain samples.

If you look at Kilauea on the island of Hawai'i, from afar you'll notice its low-lying, very wide 'shield' shape, with no visible steep-sided volcanic dome on top. We know it's a volcano because it erupts bubbling rock, and therefore it fits the description above. But unless you are standing next to the active volcano itself, you might not even realise you were on the side of a volcano, as the ground is so flat. However, if you have the urge to hike up a volcano and don't like a steep climb, then perhaps choose a shield volcano like Kilauea for an easier walk.* The reason these

---

* A shield volcano is characterised by its low profile, with gently sloping sides, resembling a warrior's shield on the ground.

volcanoes have such shallow slopes is because the basaltic lava that produces them is very runny, lacking the stickiness to pile up and form a steep slope. Nevertheless, shield volcanoes can still be very high, so if you do manage to reach the summit, you'll still get a good view. Mauna Loa, the volcano neighbouring Kilauea on the same island, is over 4,000 metres (nearly 14,000 feet) high and is considered the largest active volcano by mass and volume on Earth. Yet while shield volcanoes like these are the most common type in the Solar System, they are not the most common type on Earth. Those formed by subduction zones[*] and other volcanoes that make up the vast mid-ocean ridges are much more plentiful, all a result of the motions of plate tectonics. But such plate motions do not occur on other planets or moons, as far as we know, so looking at Earth's shield volcanoes, formed without plate tectonics, is certainly a good place to start if we want to learn more about extraterrestrial volcanoes.

If you want to appreciate the true extent of a shield volcano, they are best viewed from afar, perhaps even from orbit. Indeed, this is how we have always looked at Mars' Olympus Mons, arguably the most famous of the Solar System's volcanoes, and also a shield volcano. Such volcanoes are abundant throughout the Solar System, and they are easily recognisable thanks to their similarities to our own.

Yet there are many other landforms and features on other planets and moons that don't look like 'classic' volcanoes, but still seem to display volcanic activity. For example, we have found plumes shooting into space from

---

[*] Subduction zones are where two lithospheric plates move towards one another and one 'dives' below the other, being 'subducted' into the Earth's mantle.

the moon Enceladus in orbit around Saturn, as I've mentioned in the Preface, but there are no signs of volcanic domes on its surface. And in other places we see the evidence of icy material that appears to have flowed across a world's surface just like lava would on Earth. But can we call it lava if it's made of liquid ammonia and produced by a volcano made of ice?

The question is: how do we define what a volcano is if they all look and behave differently, with some not even producing a mountain on the surface? Consequently, what constitutes a volcano is a contentious issue even in the scientific discipline of volcanology. So much so that in 2010, a group of scientists met at a conference in Acapulco, Mexico, to discuss exactly this issue: what is a volcano? They had noticed that their science had progressed in recent years, thanks in part to the fast developments in space exploration since the 1960s. In 1989, the NASA Voyager spacecraft had captured images of the geologically young and rugged surface of Neptune's moon, Triton, with features that appeared 'volcanic' in origin. The weird thing was that it didn't appear to be erupting hot rock magma, such was its huge distance from the Sun. Instead, these 'volcanoes' might have been outpourings of water or other ices. In our description of a volcano above, as a rupture in the crust where hot liquid rock bubbles up, Triton's features were not 'volcanic'.

What the conference scientists discovered, to their surprise, was that there was no consensus on the definition of a volcano, even amongst themselves. Some definitions centred on magma rising, others on eruptions, and others on the edifice that was created in the process.

The study of volcanology began on Earth because that is where humans live. As a result, many scientists working in the field have attempted to define what a volcano is using their Earthly knowledge and what they've observed here.

Unfortunately – or fortunately, depending on how you want to look at it – the Solar System's planets and moons don't all adhere to the same rules, which certainly has the effect of keeping things interesting, but also poses some problems. It doesn't mean that our volcanologists were necessarily wrong, but their definition of a volcano might have just been too specific. They needed to increase their sample size, not just of volcanoes, but of planets and moons too. When we look at the Solar System as a whole, there is so much variation that it forces us to rethink what is normal. Just because something was first studied on Earth doesn't make it 'normal'. Perhaps Earth is the weird one?

We could just define Solar System volcanoes as locations where fluid from the interior of a planet or moon is erupted. That's quite a broad definition and could be more useful for encompassing the varied swathe of volcanic-like features that have been observed. You'll see, however, that I haven't mentioned magma or lava here, and that's because molten rock might not necessarily be present in all these settings. We must keep the definition as general as possible because of the variation that's already been observed. Magma on Earth is molten rock that becomes solid at atmospheric surface temperatures. Extraterrestial volcanism might be characterised by geyser-type eruptions, where plumes of water or methane are forcefully ejected instead of molten rock. The release of plumes of material from a planet or moon represents the method by which that particular cosmic object has chosen to cool itself. In these cases, it is not molten rock, but other substances that form the volcano. We could simply call these 'erupted materials' to encompass all the different types of non-rocky volcanoes. However, the scientists who met in Acapulco concluded that the term 'magma' · could be used to describe all these materials. Magma doesn't just have to apply to molten rock, but can describe any fluid associated with volcanoes on other planets or moons. The fluids and gases found in these locations *are* their magma. After all, the surfaces of other planets and moons are not necessarily made of rock, but sometimes of ice. In these instances, it makes sense to call liquid versions of their surface material 'magma', just as on Earth.

The title of this book came about for two reasons. On Earth, volcanoes can represent the physical meeting of 'fire' (to represent red-hot rock) and ice. Our volcanoes are always hot, even those erupting below freezing cold glaciers in Iceland, or at the base of the frigid oceans. Volcanoes can erupt exceptionally hot material into very cold settings. However, while such settings on Earth, or elsewhere, could also represent the 'ice' part of this book, I am also using the word to include the Solar System volcanoes that are not hot: those that, instead, erupt ice itself, which are known as cryovolcanoes. As we've just seen, the problem comes in defining such settings as volcanoes, and has historically been a contentious issue. These ice volcanoes don't necessarily look or act much like volcanoes on Earth, but that doesn't mean they're not volcanic in nature. The important part is that volcanic activity, wherever we look, must be fuelled by energy contained within the object, which provides heat to move materials around. On the inner planets and Io, this energy is enough to melt rock. But on the icy worlds less heat is needed to get their insides moving, since the materials they are made from freeze at lower temperatures. Hence, as we shall see again and again, active volcanoes are a sure sign that a planetary object is alive inside.

## Earthly transformations

In human timescales, Earth's large-scale tectonic changes are not so apparent on the surface. On a small scale we can observe some changes, such as the formation of a new portion of land in a matter of weeks when a volcano produces an exceptionally large burst of lava. These changes are the length of a few kilometres at most, so unless you live very close to where the lava has flowed, you're not affected. It is when we view these changes over millennia that they become more significant. Entire continents are formed and move their position around the globe. New islands that develop as underwater seamounts anchored to the seafloor emerge out of oceans and can eventually grow to have snowy summits, while still being connected to the Earth's surface kilometres below the ocean. Landforms can even completely disappear as they are recycled back into the Earth's interior, slowly but steadily destroyed by erosion, or covered over by lava flows, and thus are lost at the surface, being subsumed into the geological record. These changes all form part of our planet's rock cycle, creating igneous, sedimentary and metamorphic rocks.

It is the process of plate tectonics that is responsible for these mostly gradual, yet continual, Earthly transformations. Importantly, without the cyclical transformation of rocks on Earth, our planet wouldn't be the place it is, capable of housing such a varied collection of plants and animals.

Simply put, the outer shell of the Earth is unstable. This part of our planet is not just made up of the crust, but also includes a portion of the upper mantle, which sits just below the crust. Together this is called the lithosphere, and it is a solid yet brittle portion of the planet, broken up into a multitude of irregularly shaped pieces of varying sizes like a very large, planetary-sized jigsaw. The jigsaw pieces are called tectonic (lithospheric) plates and there are seven major ones, and dozens of smaller ones. The largest plate forms the base of the Pacific Ocean and is called, not unsurprisingly, the Pacific Plate.

The theory of 'plate tectonics' accounts for the movements of these plates in relation to each other, and they are often described as gliding about atop the Earth's 'gooey' mantle, which sits below. Scientifically, this description is not completely accurate, but it does a good enough job of representing the situation simply enough for us to imagine it.

The interior of the Earth is made up mostly of its upper and lower mantle, extending from around 10 to 70 kilometres (6 to 45 miles) in depth (depending on how thick the crust is in that location) down to 2,900 kilometres (1,800 miles), where the lower mantle reaches the Earth's outer core. Yet, despite its volume, it is a very misunderstood region of our planet. The mantle is often thought of as a molten sea of magma, but this is a misrepresentation, as it is, in fact, nearly 100 per cent solid. The problem is that parts of the mantle behave in a way that make us think of it more as a liquid that flows, or convects,* as it moves heat around from the Earth's interior. This flow can be imagined as a creeping action, moving very slowly as the individual microscopic grains of mineral crystals that form the mantle are stretched out under the enormous pressures within the Earth. I like to think of it as being like tarmac on the roads, which can sometimes become a little bit squashed under the pressure of cars during an especially hot summer. The tarmac is a solid material, but when it becomes hot, it can move and behave more like a very viscous liquid. Thinking of the mantle as a set of interconnected mineral grains that can stretch and move in relation to one another helps us to picture this region of our planet as solid, while explaining

---

* Convection is the transfer of heat due to the bulk movement of molecules, which causes the material to flow.

how it can 'flow'. This may sound like an inefficient process but over geological time, it is involved in producing impressive, and sometimes terrifying, results – mountain ranges, volcanoes and earthquakes – as it plays its part in the process of plate tectonic movements. All of these features result from the fact that the planet is cooling down and finding a way to redistribute its internal heat.

The lithospheric plates that sit on top of the mantle come in two varieties, composed of either oceanic or continental crust, which are connected to the very upper portion of the upper mantle. We will explore these in more detail later. This part of the mantle doesn't behave anything like a liquid; unlike the majority of the mantle below that extends to the core, the upper part is just as solid and brittle as the crust it is connected to at the surface. However, the block of rigid lithosphere sits on the so-called asthenosphere, which is also a part of the upper mantle. 'Astheno' means 'without strength' in ancient Greek, which helps to describe this layer within Earth. It plays an important role as it behaves in a weak, plastic manner, being able to deform easily so that it can convect and flow. Despite this movement, most of the asthenosphere is still not hot enough to actually melt the rock it contains, yet the warmer parts of it do gradually flow upwards and the cooler parts flow down, creating convection cells.

As I've already said, although we can wildly oversimplify this scenario by saying that the tectonic plates 'float' on the slightly squishy asthenospheric layer below, we shouldn't imagine them as inflatables serenely bobbing around in a swimming pool. Their journeys are not always smooth and wherever they meet, there is geological activity, some of it rather alarming. Since they are all connected, the movement of one plate affects another in a cascade, and the resulting activity, whether it is marked by volcanoes or earthquakes, or both, is very much dependent on the setting. In some places, plates may move away from each other serenely, but at the other end of that plate it is a very different story. When plates travel towards one another and their boundaries push up against each other, the result is anything but graceful. The lithospheric rock forming the plates is either crumpled up to form mountains, or drawn back down into the mantle in a process known as subduction. Such activity may sound alarming, but without these tectonic processes we wouldn't have Mount Everest as part of the impressive Himalayan mountain range in Asia, or volcanoes such as Mount Fuji in Japan, Mount St. Helens in the Pacific north-west of the USA, or many of the famous and beautiful explosive mountains along the 'Ring of Fire' surrounding the Pacific.

The tectonic plates of Earth form an immensely complex system. The individual plate boundaries consist of a jumble of individual faults and/or volcanoes. The overall system can be understood very broadly – one plate moves towards or away from another – but to get to grips with the fine detail at the level of individual faults or volcanoes takes much more effort. As a result, scientists are still trying to provide meaningful and accurate forecasts of when earthquakes will strike, or when volcanoes will erupt, as they need to understand an enormous amount about the different mechanisms that make up the overall system.

Putting the hugely destructive earthquakes aside, plate tectonics have many impressive features. One is that the constant movement of Earth's outer shell produces a significant amount of geological variety. The volcanoes that result from the interactions of one tectonic plate with another differ in size, shape, chemistry and eruption style. Each separate system can produce many volcanoes, and not one of them is exactly like any other. Just because one volcano is situated close to another, even if both evolved for the same reason because of the exact same geological setting, there is no guarantee that they will be alike. It doesn't mean that they will erupt at the same time, which means that the forecasting of eruptions, for obvious reasons, is a major and important component of volcano studies. Understanding just one volcano in detail – the reasons how and why it erupts – can take a human lifetime of research.

Volcanoes produce a variety of different landforms when they erupt. Part of the science of trying to figure out what is going on *beneath* the surface is to study their appearance *at* the surface, which is all we are able to achieve with the vast majority of volcanoes on other planets in our Solar System. The surfaces of volcanoes and their flows can be rough or smooth, they can be different colours, and the volcano itself can have very steep or shallow sides. These are all features that can help a scientist determine what the volcano is made from, and how it formed, even before collecting a sample from its slopes.

## Looking out to space

Applying our knowledge about Earth's volcanoes to the many existing in space has its limitations. Not only are volcanoes in space hard to get to, and difficult to obtain images of and measure and analyse, but their eruptions occur in very different environments to those on Earth. Nevertheless, the best way for us to understand what we see in space is still to study Earth's volcanoes. If we can understand what is going on inside our own planet, we can start to understand what might be happening elsewhere in space, since volcanism is the surface expression of what is taking place deeper underground.

Because of the complexities of studying other planetary bodies, we must begin rather simply when we want to

examine extraterrestrial volcanoes. Scientists investigate
the volcanoes themselves by looking at their location
and appearance, mostly in images taken from fly-by
missions, and in some cases from orbiters. Yet they must
also understand the alien conditions that make up the
environments surrounding these fascinating features. I'm
talking here about factors like crushing pressures, vacuum
or extremes in temperature, to mention just a few. They
not only affect how a volcano erupts, but what happens to
the material, or magma, it brings to the surface.

Nevertheless, if scientists are to understand the inner
workings of planets, then looking at what happens on the
surface is a great place to begin. It's a case of planetary-
scale bookkeeping, documenting past events in order to
piece together their history. Active planetary bodies
literally spew out their insides, which offers a glimpse of
what they contain. Scientists just need to be mindful of
how the seemingly bizarre environments of these foreign
worlds affect the volcanoes they host, and the materials
they produce.

As we've seen, this branch of science, comparative
planetology, involves comparing the natural processes that
occur on different planetary bodies to help in the
understanding of others. Although volcanism is often
studied in isolation, it is joined by atmospheric science,
geology, hydrology and possibly even biology (or
astrobiology when it refers to places other than Earth) to
name just a few. It is essential we understand the whole
system and the complex interplays that occur between
geological and other systems.

You may wonder how studying other planets helps
scientists to learn anything about Earth. Surely, with Earth
being so accessible to humans, we should know more about
it than any other planet in the Solar System. While this is

true, we certainly don't understand everything yet. As I'm sure most people are aware, Earth is an unbelievably complicated entity. So, it is helpful to look outside our special blue planet at other parts of the Solar System, even if it's just to increase our sample size for scientific accuracy. For example, climate models on Earth are complicated partly because humans have been interfering with the climate for quite some time. As a consequence, trying to model and predict how our climate will change naturally over time, before unpicking the effects humans continue to have, can be aided by studying the climates of other planets. Scientists can test their models elsewhere to see if they accurately predict the climate that is observed in these places. They can then refine these models to see how the influence of human activity affects them. Volcanology is no different. By studying the large variety of Solar System volcanoes, scientists can learn more about Earth-bound ones, and even how volcanoes affect things like climate.

## Forming extraterrestrial volcanoes

One of the major differences between volcanoes on other planets and those on Earth is that the extraterrestrial ones are probably not the result of geological activity at plate boundaries. Scientists are still not certain if other planets in our Solar System have now, or have ever had, plate tectonics. However, the vast majority of evidence points towards Earth being the only planet in our Solar System with active plate tectonics.

That is not to say other planets are not active, however, and it isn't just volcanic eruptions that make them active, but earthquakes too. Or maybe, to be more accurate, we should call them moonquakes, marsquakes or venusquakes?

You may be wondering how there are volcanoes in space if there are no plate boundaries where they can form. Although I've said that volcanoes tend to appear at plate boundaries on Earth, there are plenty that appear on the surface of our planet and are totally unrelated to the action taking place at the edges of the plates. Some volcanoes can be found right in the middle of a plate, many miles from an active boundary. It is these volcanoes that might be useful to scientists studying volcanoes in space, since they don't require plate tectonics to form.

Let's take the Hawaiian Islands as an example of one of these special volcanic sites. This set of islands sits in the middle of the Pacific Plate – the largest of them all – miles and miles from any of its boundaries with other plates. The reason the islands have appeared in this location is because the volcanoes are fuelled by an immense column of abnormally hot, chemically distinct magma that rises through the Earth's mantle. This rising rock can be thought of a bit like a large chimney within the Earth, known as a mantle plume. When such a chimney of hot rock reaches the surface, it can erupt huge volumes of lava. We can study these large eruptions in the geological record, where thick units of lava flows are stacked progressively on top of one another. They are calculated to have erupted rapidly over the course of a short timescale, geologically speaking, sometimes in just tens of thousands of years. They are sometimes referred to as 'large igneous provinces', but we can also refer to them as hotspots, to represent the focused region and the eruption of hotter than normal mantle material. We'll be meeting these regions again later.

There are many older examples on Earth to discuss later too, but arguably the most famous example active in modern times is the island of Hawai'i. Nevertheless, for some unknown reason, we don't see very large eruptions there at present and all of our largest igneous provinces, which represent especially large outpourings of lava, occurred millions of years ago. Even the island of Hawai'i, despite housing the Kilauea volcano – one of the most active on Earth – produces far less lava today than during the historic eruptions.

When a mantle plume rises up to meet a large tectonic plate, it begins to form a volcano. While the plate continues to move, as all plates do, the mantle plume remains rooted deeply in the Earth. This means that the erupting lava produced by the plume punches through the crust over time (millions of years) to produce a chain of volcanoes as the plate continues to move over the stationary mantle plume. You can imagine this a bit like holding a pen below a piece of paper and moving the paper above it to leave an ink track on the underside. That ink track would show you the trail of volcanoes left behind at the surface after millennia. Of the Hawaiian Islands, Kauai is the most north-westerly, and older than Oahu sitting to the south-east, and they are both older than the 'Big Island' (Hawai'i), sitting further south-east again than both of those islands. The islands, and the volcanoes that form them, decrease in age as you travel south-eastwardly, which demonstrates how the plate is moving in a north-westerly direction. A volcano and its host island will eventually stop growing because the plate has gradually moved away from the plume source below so that a new volcano starts to form next door. Relatively speaking, the Hawaiian Islands are all young, with the oldest being just over five million years, and the youngest a matter of a few centuries old at most, and in some cases just days old, as they are still erupting and growing.

However, the big difference in space is that if a mantle plume rises up through a planet to erupt at the surface, then the lack of tectonic plate movements means that the thickness of erupted lava will grow and grow on the surface, piling up to form an enormous volcano. Olympus Mons volcano on Mars stands at nearly 22,000 metres (72,000 feet) high, making it the largest mountain in the Solar System, because this was exactly how it formed. It had two billion years to grow because there were no plates to move it away from the hotspot. Such volcanoes, whether they are on Mars, Earth or elsewhere, are still known as shield volcanoes as they are all characterised by low profiles. This doesn't mean they're not tall, but that in order to be tall they must also be very wide.

In terms of size, Earth's shield volcanoes don't compare to those on Mars. Mauna Loa, the largest volcano on Hawai'i, is just 4,207 metres (13,802 feet) above sea level. Even if Mauna Loa is measured from its true base on the ocean floor, it is still only 10,000 metres (33,000 feet) high (we have to remember that the Hawaiian volcanoes started growing on the seafloor before rising high enough to poke their heads above the surface of the ocean). While Mauna Loa's peak makes the island of Hawai'i the second highest island on Earth, it is dwarfed by Olympus Mons, which at 25,000 metres (82,000 feet) is nearly three times higher than Mount Everest (8,849 metres or 29,032 feet).

Interestingly, if a shield volcano the size of Olympus Mons attempted to grow on Earth then it would collapse under its own weight. The pressure it exerted on its base would be so great that it would literally melt the rock underneath. In comparison, the gravity on Mars is roughly 2.7 times weaker than on Earth, which means that there is less pressure pulling down the mass of a mountain, so there is less pressure acting at its base. This means Olympus Mons can keep growing and growing … until it reaches that critical height where Mars' gravity means that it can grow no more. That point may now have been reached: activity at Olympus Mons appears to have stopped, but volcanic activity on Mars may well have dried up for other reasons too.

As we've seen here, tectonic plates are not necessary for volcanoes to form on a planet or moon, and a lack of plate tectonics means that volcanoes have the ability to grow very large indeed. Inevitably, the presence of plate tectonics creates variety. It can make all types of volcanoes – from steep-sided, snow-capped mountains to small seamounts on the seafloor. The main thing we need for a Solar System body to be volcanically active is internal heat, which the body needs to lose. I'll discuss the ways in which planets and moons warm up and cool down in more detail in Chapters 7 and 8, because there are actually many different ways, and these, coupled with their size and location within the Solar System, control the type and extent of volcanism that appears on their surfaces.

**Fire and ice**

You have probably realised by now that the Solar System is littered with volcanoes, but that they aren't necessarily like those on Earth. In fact, while volcanoes represent some of the hottest places on Earth, they are also found at some of the coldest places in the Solar System. Many cosmic volcanoes are not even made of magma, but ice!

Again, I must stress the importance of the ground-breaking NASA Voyager missions of the 1970s and 1980s, particularly in terms of the increase in knowledge about objects in the outer Solar System. The images that were beamed back from space forced scientists to think differently about our Solar System, and to re-think what constitutes an 'active' planet. The Voyager spacecraft captured images of unanticipated activity in unexpected locations throughout the Solar System. For example, Triton's surface looks young because it is currently active, erupting plumes of gas, dust and ice particles kilometres high. Such 'geyser-like' eruptions have been termed 'ice volcanoes' and are otherwise known as 'cryovolcanism'. There is no parallel on Earth for these eruptions but that doesn't mean they aren't a common type of volcanic activity in the Solar System, and possibly beyond; they just don't look like anything we've encountered on Earth before.

More recently, the New Horizons mission to Pluto, which launched in 2006 and arrived at Pluto in 2015, revealed that this apparently cold and dormant 'would-be' planet might have volcanoes made of ice. Such volcanoes even appeared to have been active in the geologically recent past, which was particularly unexpected as Pluto was thought to be too far from the Sun, and therefore too cold, to be active. What scientists have found out in the past few decades is that volcanism in the Solar System is enormously varied and it is known to occur, or have occurred, on the surface of many of the planets and moons out there, with ice volcanoes being just as common as those made of fire.

Of course, volcanoes – even the hot ones – aren't actually on fire. I'm using the term 'fire' in this book because molten rock glows like fire. The lava spewing from a volcano on Earth can reach up to 1,250°C (2,282°F), a temperature not dissimilar to a wood fire. Fires are hot because they represent the sudden release of thermal energy and light. When a fire starts, initially bonds are broken in a process known as pyrolysis, which releases individual atoms and energy. These highly reactive atoms don't stay single for long, though; they soon bond with oxygen, of which there is plenty hanging around in the air on Earth, in a process known as oxidation. A fire continues to burn for as long as it is hot enough to continue this atomic shuffling in a type of chain reaction.

Despite people often describing something as 'red hot', the hottest flames are actually blue, a colour more commonly associated with something cold, such as ice; but it is also the colour usually seen at the very hot base of a flame, which can be 1,400°C (2,552°F). If you need a really hot blue flame then you'll need to burn a hydrocarbon called acetylene, whose flame can reach anything up to 3,000°C (5,432°F) and can therefore be used for underwater welding. As they cool down, flames turn white, orange, then red, colours that we are more used to associating with hot objects such as candles, which are, on average, around 1,000°C (1,832°F).

If we imagine a fire on another planet, say Mars, whose atmosphere contains less oxygen, then we can take a guess that it wouldn't burn for as long, as a fire needs oxygen to keep going, along with a continued supply of fuel and heat. Interestingly, if a flame burns in a microgravity environment, then the flame shape is different, squatter rather than elongated, and it burns less vigorously because of the way in which the lack of gravity affects the combusted products. This has been tested on the International Space Station by brave astronauts. Rather them than me lighting a fire in an enclosed space capsule! As you can see, the space environment has wide-reaching effects that you might not have initially considered, and this is part of the reason why studying volcanoes in space has its challenges.

Lava, such as that on the Hawaiian Islands, is often also described as being red hot because, just like a flame, it is. Red hot lava is molten rock and as it cools it turns to darker red, then brown; however, it can still be over 500°C (932°F) at this stage. It eventually turns black as it transforms into a solid state. As we know, volcanoes are not actually on fire, yet the colour associated with them comes from the exact same colour-temperature scale we see for flames.

The colour of hot, glowing lava occurs because it emits, or radiates, light, in a process called incandescence. When something is glowing because it is hot, you are seeing the heat, but you see it as light. This is because the hot material is releasing some of its thermal vibration energy as photons, making electromagnetic energy.

Some very special volcanoes can glow in colours other than just red. The colour of a flame can also be affected by the substance, or the elements, being burned. Many of us will remember carrying out the flame test in school science lessons. It's an experiment I always enjoyed because the colours produced from burning different elements are so beautiful and it's a neat and easy test to do. You may recall the stunning green–blue flame of copper and the wonderful violet flame of potassium.

Some volcanoes are especially and spectacularly colourful. The lava from Indonesia's Kawah Ijen volcano glows an eerie cerulean blue, hence the nickname 'the blue volcano'. In fact, it is not the lava itself that glows blue; it is the standard red just like other volcanoes. The blue colour comes from an abundance of sulphuric gases emitted by the volcano. As the gases are exposed to oxygen in the air and combine with the heat of the lava where it erupts at the surface, they ignite and glow blue. At night, these 'lava flames' appear to flow along the side of the mountain as they move above the lava itself, producing a hypnotising effect. The flames were initially caught on camera by a journalist making a documentary to highlight the harsh working conditions experienced by sulphur miners in the region. Sulphur is not an element you want to inhale because it causes all manner of adverse health conditions and breathing difficulties. At Kawah Ijen the gases are at toxic levels, over 40 times the safe breathing level (as set by the UK). Yet sulphur is a commodity in demand as it is used in a whole host of applications from batteries, to fertilisers, to fireworks, amongst many other things. Unfortunately, people in the region knowingly, or unknowingly, ignore the risk to their lives to work on this dangerous volcano so that they can retrieve the solid yellow deposits formed when the volcanic gases cool at the surface.

What this special volcano shows us is that even when it is viewed from a safe distance, it can still tell us a lot about the chemistry of the volcano itself. Such knowledge is important to remember when we venture out to explore the Solar System, where we can't necessarily go and probe each and every volcanic vent to see what mix of gases they emit. Simply capturing images of other worlds can be very informative.

## Peering into the Solar System

While each planet is unique, inhabiting its own private location in the Solar System, they often share similarities with other planets, and sometimes with Earth too. As we have seen, studying Earth's volcanoes to understand our extraterrestrial volcanic friends can be helpful, even if at first sight they appear to be very different. While Jupiter is a gas giant that is very different from Earth – much larger and more gaseous being the most obvious distinctions – scientists have discovered similarities between Jupiter's moons and our own precious blue planet.

Let's take the moon Europa as an example. The location of this moon within Jupiter's gravitational pull, millions of kilometres away, may seem alien, but that doesn't mean it's necessarily other-worldly in its physical characteristics. Europa is thought to have an iron core, a rocky mantle and an ocean of salty, liquid water, one that holds more water than in all of Earth's oceans. The main difference here is that atop Europa's liquid ocean is a surface made of water ice which forms a crust, albeit one that is not made of rock, but scientists have figured out that this surface is actually relatively young, maybe just 40–90 million years old. This evidence comes from Europa's cratering record. By counting the number of craters in images they have of the moon, scientists can calculate the age of its surface. Simply put, the more craters there are, the older the surface is, because it has had more time to accumulate impacts of comets and asteroids from space.

Europa's young surface is thought to be tectonically active, which means it is capable of repaving itself, effectively destroying those craters. If this is the case, then Europa is likely to be warm inside, and, therefore, it has the potential for volcanic activity. In 2016, the Hubble Space Telescope captured intriguing images of potential ice plumes emanating from Europa's surface. Although not yet proof of cryovolcanism, these bring us a few steps closer to finding out. The Europa Clipper mission will launch in the 2020s to take a closer look at this fascinating moon. If the plumes are proven to be the result of volcanic activity, then all these features point heavily towards Europa being capable of supporting life too.

One of the interesting aspects of studying planetary science is that scientists try to take the ideas and concepts they use for Earth and, as we have seen, extrapolate them to other planetary bodies. They have only, for the most part, a few images of parts of their alien surfaces to work with, and yet scientists often have a good grasp of the basic conditions existing on the surface of a planet or moon: its composition, temperature fluctuations, pressure, gravity and atmosphere. However, the challenge is that they need to take calculated guesses as to how volcanic products might emerge onto, and react with, the seemingly strange environments they meet. It is only when they can send a complex and sophisticated space mission to visit these places that they can observe them in detail and see if they were right. For example, we have seen that Earth and Mars share some volcanic similarities, namely shield volcanoes. The difference is, despite Mars and Earth both having substantial gravity and an atmosphere (even if Mars' is very thin), as we saw, the surface conditions on Mars allow its shield volcanoes to grow to much larger sizes.

What if we were to compare Earth's eruptions to those in places such as the Moon, or perhaps another of Jupiter's moons, where gravity is so much lower? For one thing, we could expect eruptions to be much more explosive; for another, pyroclastic flows can't function without an atmosphere, and without wind they couldn't carry volcanic ash away.

What if we went to Venus instead? There the gravity is fairly similar to that on Earth, but Venus' atmospheric pressure is 100 times higher, and its surface is considerably hotter (at around 480°C or 896°F). The erupting volcanoes on Venus don't pop and explode, they ooze, and because of the high surface temperatures, any surface flows cool much slower than those on Earth, meaning that they can travel a long distance.

So far, I've only mentioned a few of the many 'fire' planets (those that produce hot magma) and we've taken a brief glimpse at some of the ice planets. However, we clearly need to delve into these environments a bit further and luckily, you're in the right place.


## Life

Earth is unique for all sorts of reasons, the most obvious being that it is the only planet to host life, as far as we know. It might also be the only place in the Solar System to have, or have had, plate tectonics. One question we should ask is whether these two features are related. Is it necessary to have plate tectonics for life to develop on a planetary body? Answering the question of whether we're alone in the Solar System is not a simple one. We *appear* to be the only planet capable of hosting intelligent life, but that doesn't mean other forms of life don't exist out there, even if they are now extinct or not capable of making themselves obvious to us. Perhaps plate tectonics are not important for life, but it is likely that a planet needs to be active, inside and out, for life to take hold and survive for any substantial period of time.

The thing we need to get our heads around, however, is that not all life resembles human life. While we depend on an agreeably warm, wet and well-oxygenated environment to survive, there are many organisms on Earth that don't share our needs. We've discovered life on Earth that enjoys, even thrives, in the weirdest environments imaginable, such as the crushingly high pressure and cold depths of the oceans, where no sunlight can reach. We've even discovered life in some of the driest or most acidic places on Earth, where we thought nothing could ever survive. The organisms that exist in these strange settings are sometimes known as extremophiles – they positively love extreme conditions – and they force us to open our minds, and our search window, for life in peculiar extraterrestrial environments. Just because a planetary body doesn't look, behave, or feel like Earth doesn't mean that life isn't a possibility there.

One of the reasons scientists study the distribution and type of geological activity on other Solar System objects is to figure out what is going on below their surfaces. What is the driving force deep within their interiors? Why are they are warm and how do they cool themselves? Each is unique, and without studying them in detail to understand their distinctive Solar System environments, scientists have no hope of figuring out whether or not they are still active. In fact, only when they have truly understood the present and past environments of these objects can they begin to answer the 64-million-dollar question: is there life elsewhere in the Solar System?

It might be the case that plate tectonics are not necessary for life but that volcanoes are. Volcanoes are almost certainly essential elements in the delicate balance of our own planetary environment. Without them, Earth wouldn't be the functioning, life-giving planet that it is. However, that is not to say they are vital for providing a life-giving environment everywhere in the Solar System. Whether or not volcanoes are necessary for life is a research question that remains unanswered for the time being.

Whatever the case, there is tantalising evidence to suggest that volcanoes played an important role in allowing life the opportunity to get a real foothold on Earth. Scientists know that around 2.5 billion years ago, Earth began seriously accumulating oxygen in its atmosphere. Oxygen initially appeared thanks to the rise of oxygen-producing cyanobacteria, but the geological record shows them to have appeared 200 million years earlier than this dramatic change in the atmosphere. If the bacteria were responsible for the oxygenation of the planet, they were a little sluggish to start with. So, why did their oxygen production suddenly peak? It was then that scientists noticed that around the same time as atmospheric oxygen started accumulating with gusto, there was a sharp change in the style of the planet's volcanism, from undersea to subaerial, or terrestrial, eruptions.

While this might not seem instantly important, the key fact is that undersea eruptions are known to be effective at scrubbing, or cleaning, oxygen out of the atmosphere. They produce more 'reducing' gases that remove oxygen by bonding with it. This meant that those early oxygen-producing bacteria were effectively fighting with the volcanoes. However, around the same time, the continents had started to thicken due to plate tectonic movements, meaning more and more land was appearing above sea level. Researchers say this shift in eruptions taking place on solid ground, as opposed to underwater, coincided with the increase in atmospheric oxygen. Terrestrial volcanoes are known to be less reducing than their undersea counterparts, meaning they soak up less oxygen. This shift in the balance of volcanic activity on Earth allowed for the oxygen produced by the bacteria to begin accumulating in the atmosphere, a consequence which undeniably gave rise to a whole host of animals making Earth their home.

If the interpretation of these observations holds true, it is because of a relatively small shift in global events that eventually led to big changes for our planet: the difference between life being given a kick start or struggling to ever get going. This is a sobering thought at present, when humans are responsible for altering Earth's atmosphere at such an unprecedented, and unnatural, rate. It didn't take the Earth long to create the right conditions for life to really take off, so how can we expect it to maintain those conditions when we are creating such dramatic global atmospheric changes now? It is a fine balance and one that will hopefully stay tipped in life's favour, but only time will tell.

While volcanoes might have been responsible for helping to start and maintain life on Earth, they have had a tendency to end life too. I will explore this topic in the following chapter. Understanding the very nature of volcanoes is critical if we are to forecast how they will behave in the future, and therefore be able to prevent major loss of life.

# CHAPTER TWO

# Destruction

A round a tenth of humans live within the potential reach of hazards posed by volcanoes. You might think that if you can't see a volcano, then you're safe, but eruptions can have local, regional and even global effects.

The hazards posed by volcanoes are important phenomena for research partly because the results of volcanic eruptions affect such a huge portion of the planet's population. Despite this, the majority of volcano-related deaths are usually in the relatively local environment; a region a few tens of kilometres from the summit is statistically the riskiest area. There is a size–frequency relationship where we find that larger volcanic eruptions are less frequent, but they understandably have a larger risk radius. Nevertheless, the dangers from small eruptions can still be high, if not as far-reaching geographically.

There are some 1,500 active volcanoes on our planet, with 'active' being defined as a volcano that has had suspected or confirmed activity within the last 10,000 years. The fact that over 29 million people worldwide live within just 10 kilometres (6 miles) of these active volcanoes, and that 800 million live within 100 kilometres (60 miles), are worrying statistics.

Sure enough, volcanoes often get a bad press, but as the cause of major natural disasters that devastate swathes of land, destroying property and killing people, it's no surprise. The problem is that there is an astonishingly large variety of volcanoes on Earth and they all exhibit their own subtle and conspicuous signs before an eruption.

While each individual volcano is broadly characterised by a certain type of activity – whether that be the production of large ash clouds or flowing lava – they can easily show both of these eruption styles, and many more. Furthermore, a volcano's 'personality', if we want to call it that, can change without much warning. Therefore, forecasting the hazards volcanoes pose to their surrounding communities is a tricky science often fraught with complications.

An important aspect of managing the hazards a volcano poses is in communicating the risk to the local population who may be required to evacuate their homes. The problem though is that scientists are not entirely sure that their forecasts will be accurate; they are simply putting together a lot of information from different datasets about the volcano's behaviour over the past few years, months and weeks, and making an educated guess as to how they think it will behave in the future. As time goes on, scientists will continually improve their forecasting skills, and there have been many examples in recent decades where they have been spot on, evacuating entire towns just in time. However, in the rare event that they are wrong, and the volcano decides not to erupt, they run a tricky gauntlet while trying to maintain the trust of the public. After all, no one wants to abandon their home; and if it is an unnecessary evacuation following a false alarm, then people are understandably annoyed, and they lose confidence in the future warnings. But this situation is now relatively uncommon as the science of hazard forecasting is constantly and rapidly improving.

However, it's not just the local population that might be affected during an eruption. There is a whole raft of hazardous after-effects of volcanic eruptions, some of which are less well known, but no less deadly, with sometimes far-reaching – and we're talking global – effects.

When it comes to other planets, the situation is no different, except that they obviously pose no risk to life since there is apparently none there. However, volcanic eruptions on other planetary bodies can have significant global effects on their atmospheres and surfaces. Scientists have studied extraterrestrial eruptions by looking at their deposits from afar, generally after the eruption has ceased, using spacecraft images and data. The next step is usually to compare what they see to volcanic features that occur on Earth and have previously been studied. Using comparative planetology, scientists can learn more about the features in space that they can't physically go and sample. But as we've already discussed, it works both ways: sometimes these extraterrestrial volcanoes can help us learn more about volcanoes on Earth.

## Flowing rock

Despite lava flows being arguably the most well-known type of volcanic activity, whether the lava pours, sprays or is squeezed out of a volcano, it is not usually very dangerous. While it can on occasion maim or even kill, most people could outrun, or even outwalk, most lava flows. Flowing lava tends to creep sedately over volcanic terrain. It can, however, become contained within a lava tube, which forms when lava flows through a natural conduit where the outer edge of flowing lava has cooled to form a crust, but molten lava continues moving through the middle, insulated by the crust. If this happens on a downhill slope with you at the bottom, you'd probably need to break into a fast run to stay ahead of it, but you could also sidestep the flow to avoid being caught in its fiery clutches.

Having said that, there are rare occasions where lava has been known to flow at speeds of up to 60kph (40mph).

One of the most famous examples was the 1977 eruption of the Nyiragongo volcano in the Congo. During this eruption a fissure suddenly opened up and released molten rock stored in the volcano's lava lake, which had been boiling away since 1894. This breach acted like a dam bursting, sending lava flooding down the flanks of the volcano and inundating villages on the lower slopes, resulting in the deaths of 70 people. However terrifying this might appear, it is unusual. One of the reasons for the very fast-moving lava was that this is a steep-sided volcano, but it is especially rare that this type of volcano would have a large volume of liquid lava close to its summit.

On Earth, while molten rock flows are not particularly life-threatening, they can consume entire buildings as they creep sedately along the ground, literally encasing everything in their way. This was evidenced quite clearly during the fissure eruptions from the Kilauea volcano on the island of Hawai'i that made world news in 2018. While the film content coming from the island was incredible to watch, over 50 buildings were destroyed by the lava. Thanks to great work by scientists, the local population were evacuated beforehand, but the devastation was still sorely felt.

In many cases, the bigger problem with lava flows is their release of invisible toxic gases. Most notably, sulphur dioxide can cause serious breathing difficulties, as well as eye and throat irritation, for anyone close by. Because of this, the advice is always to stay away from active lava flows because even if the lava itself doesn't cause a problem, the gases released might.

## Bang, fizzle, pop

While lava flows don't seem to cause much danger for humans on Earth, many volcanoes have the potential to produce more explosive eruptions, which pose more of a risk. Explosive eruptions tend to produce pyroclastic materials. The term pyroclastic stems from the Greek for 'fire' and 'broken', indicating that we are dealing with exploded pieces of hot rock. Pyroclastic materials are any volcanic fragment hurled out of a volcano. This can be anything from a boulder the size of a car to the finest ash. Generally, the larger the fragment, the less distance it will travel because the weight of a large rock will prevent it going too far from the volcanic vent, despite the often supersonic speed at which they are ejected. The smaller-sized fragments of ash can reach much further and, in some ways, are even more dangerous.

Most of us know well the story of the Roman city of Pompeii, having learnt about it at school, and maybe even having visited the archaeological site itself. I've been there myself, in the middle of summer, and it was hot. Obviously, the temperatures I experienced as I sweated my way around the excavated ruins were nothing compared to the conditions its inhabitants experienced when Vesuvius erupted with unthinkable consequences back in AD 79. It wasn't just Pompeii that was buried under 4 to 6 metres of incandescent ash and pumice; so too was the town of Herculaneum and many surrounding villas.

The eruption that caused all this destruction was immense and complicated, estimated to have measured 5 on the Volcanic Explosivity Index (VEI), making it similar in size to that of the Mount St. Helens eruption in 1980. VEI is a relative measure of the explosiveness of volcanic eruptions, devised in 1982. The scale goes from 0 to 8, with the lower end marking the non-explosive eruptions and the upper end representing a supervolcanic eruption, which we'll come on to shortly.

In terms of the Vesuvius eruption of AD 79, despite it being nearly two millennia ago and within the first century of the Roman Empire, we have a wealth of information from eyewitness accounts, most notably from Pliny the Younger. He observed the eruption from a relatively safe location around 30 kilometres (18 miles) away in the town of Misenum (present-day Miseno) and in recognition of his work, this type of eruption is now called 'Plinian'. Pliny the Younger wasn't completely untouched by the eruption, as Misenum was covered in ash and experienced devastating earthquakes throughout. Nevertheless, what Pliny observed tells us a lot about how the eruption developed. Earthquakes were recorded in the region up to 16 years prior to the main eruption, but unfortunately these weren't understood to be warnings of unrest at the volcano. I expect residents just got used to earthquakes, never thinking they indicated that the mountain looming over their towns and cities was beginning to awaken with major historical significance.

The start of the eruption was first noticed on 24 August, AD 79, when a large ash cloud appeared above the volcano. The cloud is now estimated to have risen to 32 kilometres (20 miles or 100,000 feet), which is three times the height at which passenger airliners fly. This was the Plinian part, characterised by an eruption column that ejected gas, at several metres per second, high into the stratosphere. While the cloud spread out to a more extensive head at the top, it began to rain out ash and pumice over surrounding areas, covering settlements at the base of the volcano with up to 2 metres of ash during the first 24 hours. Although raining ash might just look like snow, it is especially insidious because it is made of tiny shards of volcanic glass. If you inhale this ash, it lacerates the insides of your lungs. While this might not be enough to kill immediately, unless you already have breathing difficulties, it can produce long-term damage. Many of those living near Vesuvius will have experienced such problems a long time after the eruption.

Ashfall is all part of the tephra fall, a term that describes all the pieces of rock fragments hurled into the air by a volcano. Tephra fall is the most widespread volcanic hazard. A recent study looking at volcanic hazards estimated that, despite its wide dispersal, 80 per cent of fatal incidents related to tephra occur within 20 kilometres (12 miles) of a volcano. These deaths are often related to building collapse from the weight of deposits, whereas fatalities at larger distances are more likely to be related to the exacerbation of existing medical conditions.

In a more recent event than the Vesuvius eruption in AD 79, the largest loss of life blamed on tephra was the 1902 Santa Maria eruption in Guatemala. An estimated 2,000 people are thought to have perished owing to burial, building collapse or suffocation, or a combination of these, with deposits of over 3.5 metres (11.5 feet) in depth being reported as far away as 10 kilometres (6 miles) from the volcano.

But returning to Vesuvius, only 12 hours after the start of the AD 79 eruption, the first of the pyroclastic surges occurred. These were searing hot, fast ash flows, which proved to be the deadliest element associated with the eruption. Forget breathing in glass shards, here people were engulfed in a maelstrom of burning rock dust. There were at least four more surges that completely buried surrounding villages and towns, some of which had previously been covered in layers of fine ash. The problem with the eruption and surges is that they were both equally unexpected. Whereas raining ash does not kill you quickly, a pyroclastic surge will. They appear without warning and are instantly deadly, inundating towns and turning to stone anyone who remains.

Indeed, it has recently been reported that the brain of a young man entombed within ash in Herculaneum was turned to glass when he was inundated by the eruption, which obviously killed him instantly. He was found lying in bed, completely unaware that a flow, estimated from the analysis of charred wood nearby to be at a maximum temperature of 520°C (968°F), was going to inundate not just his house but his entire town. The extreme heat vaporised soft tissues within his body before the temperature rapidly dropped again as the flow abated, and these tissues cooled to form a glass.

Pyroclastic surges are just one type of terrifying and powerful volcanic activity that can result from a volcano collapsing, or from the ash cloud disassembling itself. Sometimes they are called pyroclastic flows, and the difference – surge or flow – simply refers to the mechanism of travel. These flows can also be classed as pyroclastic density currents, which is a term I prefer because the word 'density' reminds me that these flows are dangerous not only for their high temperatures, up to 1,000°C (1,832°F), and unimaginable speeds, up to 700kph (430mph), but they are also composed of dense rock varying from fine dust to block-sized pieces.*

Flows are denser than surges and tend to stick to the surface topography, flooding down valleys and staying on a predictable path like a gush of water. Surges, on the other hand, are a little less dense, containing more gas than rock. As such, they are able to flow upwards and over high ground and are hence probably more dangerous, as they can unexpectedly inundate higher topographical regions where people seek safety. As discussed previously, this was how the famous volcanologists Katia and Maurice Krafft were killed.

Sometimes pyroclastic currents may also contain larger blocks of rock, whether or not they formed from the collapse of the volcanic dome, and these can be called 'block and ash' flows that in fact contain very little ash and

---

* Block-sized pieces are defined as rocks over 64 millimetres (2.5 inches), but they can sometimes be the size of cars.

pumice. The force of these is certainly enough to obliterate buildings and anything else in their path.

The best way to protect people from these terrifying volcanic phenomena is to forecast the eruption and evacuate areas at risk. Of course, this relies on knowing an eruption is imminent, which isn't always simple to determine. While scientists can easily recognise that a volcano is showing signs of unrest, knowing exactly when it will erupt, on a scale of days to hours, is harder to forecast with accuracy. In recent years, volcanologists have gradually honed their skills and started to forecast eruptions more accurately, and in many cases have prevented major loss of life by successfully evacuating large areas before the volcano becomes a danger. Either way, local inhabitants will usually have had to leave their homes weeks in advance, sometimes never to return.

There are many examples I could give where scientific forecasting of eruptions in recent decades has led to successful evacuations. My favourite, partly because I have a personal connection with the island, is that of the eruption of the Soufrière Hills on the small Caribbean island of Montserrat that began in 1995. This British Overseas Territory was once a millionaire's paradise, with golf courses adorning the luscious volcanic flanks, and was also home to the famous AIR Studios owned by Sir George Martin, the record producer and composer often referred to as 'the fifth Beatle' due to his invaluable work with the band. But the problem is that the volcano forms the major southern portion of the island, which is only around 10 kilometres (6 miles) long in total. The Soufrière Hills literally loom over the capital, Plymouth, which sits on the lower flanks of the volcano. Plymouth was once home to around 6,000 people and was a bustling Caribbean town, but things were set to change in 1995 when scientists became convinced that a dangerous eruption was imminent.

Nevertheless, persuading local people who had lived in Plymouth their whole lives to leave their homes and businesses, let alone the rich and famous to desert their multi-million-dollar properties, was not an easy task. While the millionaire expats had the chance to move off the island to another of their homes elsewhere in the world, the local Montserratians weren't so fortunate. Many of them were set to end up in temporary shelters on the safer northern side of the island or, in some cases, shipped off the island completely, but not to a home of their own. Despite the major humanitarian issues involved in evacuating Plymouth, it was the right decision, because the capital was soon inundated with ash and ash flows.

Unfortunately, 19 people who did not evacuate lost their lives during the start of this massive and long-lived eruption, paying a high price for ignoring the warnings from the volcano scientists. This was a preventable loss of life and sorely felt by everyone involved in the evacuation. The situation could have been a lot worse, though. The eruption continued for years after the initial activity began, such that most of the buildings in Plymouth are now completely submerged under ash. People will most likely never return to this region of the island as it remains at high risk of further eruptions and dome collapses[*] with pyroclastic flows for decades, if not longer, to come.

I have been lucky enough to visit this beautiful, quaint and relaxed island a few times for geological fieldwork following the start of the current eruption in 1995. It always amazes me how people have coped with living next to such a beastly volcano and dealing with the hazards it poses. The island is utterly beautiful, and very lush, thanks to fertile

---

[*] A dome collapse is a partial, or complete, collapse of the volcanic edifice, or dome, that forms the volcano summit.

volcanic soil. Montserratians get on with their daily lives in the habitable northern half of the island, living with the knowledge that their volcano will continue to erupt.

## Proximity and global effects

We've already touched on this, but I wanted to reiterate that the closer you are to a volcano, the higher your risk of serious injury or death, which hopefully seems obvious. However, as we've seen, with hazards such as tephra fall, plus lahars (violent flows of mud or debris composed of volcanic material such as ash, rock and water) and tsunamis (massive waves that destroy everything in their path), people at relatively large distances from a volcano are also at risk. The apparent 'safe' distances from a volcano relate very much to the individual setting. Volcanoes are all different, with their own unique history and style of eruption.

In recent years, particularly in northern Europe after an Icelandic volcano called Eyjafjallajökull famously erupted in 2010, people have become aware that ash clouds can pose quite a risk to aircraft and, therefore, that daily life can be affected by a volcano located hundreds or even thousands of kilometres away. The ash cloud generated from this volcano ended up closing large amounts of European airspace for five days in April 2010. With airliners being unable to fly in the region during this period, there was intense travel disruption. The problem is that if a jet aircraft flies through a cloud of volcanic ash, there is a chance that the ash becomes welded to the hot turbine blades of the jet engines, rendering them seized and useless. Flying through volcanic ash at 800kph (500mph) also sand-blasts windscreens, resulting in almost complete loss of forward vision for the crew. In 1982 a jumbo jet encountered an ash cloud in Indonesian airspace at night. One by one each of the four engines failed and the aircraft descended 7.6 kilometres (25,000ft) while the crew tried desperately to restart at least some of the engines before what seemed the inevitable ditching in the Indian Ocean. Fortunately, they were successful in restarting the engines in the nick of time and landed safely at Jakarta airport, even though they could see very little through the cockpit windows.

While ash is a hindrance for aircraft – and holiday-makers – there are ways to forecast where there will be concentrations high enough to cause trouble. Because of better eruption forecasting and tracking of ash clouds, such events in the future should hopefully be avoided. However, even if you're not right next to the volcano, or in an airliner nearby, then you can still be affected by its reach.

The effect of dust moving around the atmosphere is that it can physically block the Sun, even on the other side of the world, which can cool the atmosphere. Despite this, particles of dust – particularly the larger ones – have only a short cooling effect because they are washed, or fall, out of the atmosphere within days. The problem with large Plinian eruptions is that they also eject large amounts of volcanic gases into the air that can travel much further than ash, sometimes encircling the entire globe and having dramatic and long-lasting effects on the climate.

One such invisible, yet smelly, gas that can be particularly potent is sulphur dioxide. When it's ejected in a Plinian eruption, this intoxicating gas can move up high into the stratosphere where it has the potential to combine with water to form aerosols and sulphuric acid, causing the well-known building-decaying acid rain. The tiny droplets of acid that are formed are also very effective at reflecting solar radiation, and the big problem is that the aerosols can be very long-lived, producing cooling effects one to two years after the original eruption. Perhaps though, we could view this cooling effect as a good thing, especially since Earth has been experiencing such a sharp and unnatural rise in global temperatures in recent years. Unfortunately, despite the many volcanic eruptions, they alone won't be capable of reversing these changes.

Sometimes the atmospheric cooling effects brought about by a volcanic eruption are more substantial. In some cases, after a large eruption, parts of our planet have been thrown into a so-called volcanic winter. Eruptions such as Mount Pinatubo (1991), Krakatoa (1883) and Mount Tambora (1815) all resulted in cooling of global temperatures. Pinatubo cooled the atmosphere for two to three years, Krakatoa for nearly four years, with record snowfalls recorded worldwide the following winter, and Tambora created the 'Year Without Summer'. Tambora was the most powerful eruption in human history, possibly an order of magnitude larger than Pinatubo, and with a very wide reach. The effects of the eruption resulted in midsummer frosts being recorded in New Hampshire, Maine, Vermont and northern New York in 1816, as well as deep snowfall in places such as Quebec City in June 1816. The conditions lasted for around three months, with large agricultural crop devastation in North America. The climate changes are even thought to have brought about a cholera outbreak in South Asia after disrupting Indian monsoons, which caused failed harvests and famine in the region.

Another particularly deadly eruption – in terms of its chronic cooling effects – in relatively recent times was that of Laki in Iceland in 1783–4. This eruption was not your usual bang and fizzle event; it went on for over eight months, releasing lava from 140 vents along a 23-kilometre (14.3-mile) stretch of fissures. However, the lava itself, although plentiful, was not Iceland's biggest problem.

Along with the lava there was a massive ejection of sulphur dioxide from the volcanic vents, which entered the upper troposphere and lower stratosphere, being swept into the jet stream and efficiently circling around the planet. Because of this, the northern hemisphere saw temperatures fall by about 1.3°C (2.3°F) for one to two years as the gas was turned into sulphuric acid aerosols that reflected the Sun. Acid rains were felt in Europe, with the production of acidic dews, and frosts that were bad news for arable crops. Not only that, but volcanic ash rained down as far away as Italy, over 3,000 kilometres (2,000 miles) from Laki.

But the volcano's effects were not felt nearly as badly worldwide as they were within Iceland itself. While 80 per cent of the sulphur dioxide was efficiently lofted up high during the explosive eruptions, with most of it being removed as acid rain within a few days, 20 per cent of the gas release came directly from the cooling lava flows themselves, creating a volcanic haze at the surface that stayed local, and low, over Iceland. It was this more minor component that had dire consequences on the small island. Sulphur dioxide clung closely to the ground, bringing about acid rains locally that were so strong they burnt leaves and killed off trees. This is not where Iceland's problems ended though, because the sulphur dioxide was joined by fluorine, another common volcanic gas that is not usually seen as particularly dangerous. However, the fluorine was released in such large amounts that it settled on the surrounding landscape, being incorporated into grasses and plants. Unfortunately, this was bad news for grazing livestock – over 60 per cent died from fluorosis after ingesting the fluorine-rich grasses.

While deaths of livestock on one island might not seem particularly problematic in a modern world, we have to remember that Iceland is a relatively small island that was not served by modern transport links in the late 1700s. The Laki eruption ended up killing over 10,000 people, which was over 20 per cent of the population at the time, due to famine and disease brought about by the volcanic 'haze'.

While the thought of a similar event occurring in Iceland today is still terrifying, the good thing is that, despite the potential severity, the cooling effect from volcanic eruptions tends to be short-lived – in geological timescales anyway. And the release of ash and noxious gases into the atmosphere doesn't always have to be so grim. After the 1883 Krakatoa eruption, brilliantly coloured fiery red sunsets and sunrises were seen all around the globe as the ash and aerosols scattered light of the red wavelengths. Krakatoa is not the only eruption to have had such an effect on worldwide sunsets, but it is one of the more famous ones thanks to the numerous works of art and writing it inspired. It is even thought that the volcanic sunsets brought about by the Krakatoa eruption inspired the blood red sky in the famous painting *The Scream* by Edvard Munch in Oslo. Although painted in 1893, *The Scream*, and the group of works that join it in *The Frieze of Life*, have been found to almost certainly have established origins in the preceding decades.

Munch said at the time: 'I was walking along the road with two friends – then the Sun set – all at once the sky became blood red – and I felt overcome with melancholy. I stood still and leaned against the railing, dead tired – clouds like blood and tongues of fire hung above the blue-black fjord and the city. My friends went on, and I stood alone, trembling with anxiety. I felt a great, unending scream piercing through nature.'

## Volcanic floods – it's not all about lava

Despite Pompeii's infamy, only around 2,000 people died in Vesuvius' most notorious eruption. I say 'only' because we can compare this to the eruption of Krakatoa in 1883, which killed 36,000 people. However, despite Krakatoa being the most powerful eruption in recorded human history, standing at VEI 6, the deaths of these inhabitants weren't caused directly by the volcano. Instead, they died as a result of a large tsunami, sometimes known as a mega-tsunami, triggered by the volcano.

Tsunamis have only properly entered the human psyche in the modern day. Unless you've been lucky enough to study geology or earth sciences in your lifetime, the idea of a tsunami might only have become obvious after the devastating so-called 'Boxing Day' tsunami in the Indian Ocean on 26 December 2004, depicted in the 2012 film *The Impossible*, starring Naomi Watts and Ewan McGregor. The tsunami killed around 230,000 people and was triggered by a massive earthquake occurring offshore, which rapidly displaced the seafloor and pushed a huge body of water towards the land. No volcano was involved, but it helps to illustrate the devastating nature of these natural phenomena, because tsunamis can be triggered by volcanoes too.

The concept of a volcano triggering a tsunami is less well known but clearly, as is the case with Krakatoa, no less deadly. In fact, around 5 per cent of tsunamis are triggered by volcanoes. In either case – earthquake or volcano – a tsunami just needs a displacement of material for it to be triggered. The collapse of a large portion of a volcano can displace water in the same way as an earthquake, moving portions of the seafloor relative to one another. However, even a very large tsunami wave in a coastal region may be almost unnoticeable in the deep ocean. The reason is that as the wave travels into shallower water it slows down and starts to grow in height in a process known as 'shoaling', where the amplitude of the wave increases. I use 'slows' as a relative term here. A tsunami can still hit land at over

300kph (200mph) and stand at 30 metres (98 feet) high. Just imagine a wall of water as tall as a six-storey building racing towards you. It is a terrifying thought.

Tsunamis can be an especially scary and hazardous volcanic activity, even for people who might think they are far removed, and therefore at a safe distance, from a volcano. In fact, the majority of fatal incidents at distances greater than 15 kilometres (10 miles) from a volcano are due to lahars, otherwise known as mudflows. Obviously they can affect people who reside closer to the volcano, too. These lahars are so deadly because they are extremely dense. Think of a wet concrete slurry racing downhill. They can inundate and bury entire towns at the base of a volcano. People usually die either from trauma or from drowning, but these flows can also be very hot. There is even a danger from such mudflows years after an eruption, as heavy rains can trigger secondary lahars from the volcano's slopes, much like a landslide composed of water and earth. The year 1985 saw the worst volcanic disaster of the twentieth century when 25,000 people were killed at Nevado del Ruiz, a volcano in Colombia. Despite the eruption itself being relatively small, it melted the volcano's glacial ice cap, which triggered an enormous surge of meltwater and mudflows down Nevado del Ruiz's canyons. Nevado del Ruiz truly represents a region of 'fire and ice'.

Volcanic interactions with water don't stop here. The thing is, when volcanoes meet water ice the result can be explosive and somewhat catastrophic. Obviously, the heat from a volcanic eruption can melt adjacent ice, and rapidly turn it into water. The Icelanders have given a name to floods produced when volcanoes generate large amounts of meltwater: 'jökulhlaups' (pronounced yo-kool-lahp). This, I think, is a lovely sounding word for a terrifying phenomenon, one that is not just confined to Iceland but has the potential to occur at any place where hot volcanoes interact with ice.

## Supervolcanoes

I probably can't complete this chapter without mentioning supervolcanoes. They get a lot of attention in the press, particularly Yellowstone in the USA. Thanks to the public's fascination with volcanoes like Yellowstone, and its potentially large and destructive future eruptions, the media can often report that a supervolcano is due to have a 'big one' soon, when in reality that is never the case. Nevertheless, scientists have found evidence for around 47 supereruptions from volcanoes like Yellowstone throughout Earth history, and there are even six active supervolcanoes at the present day. But just recall that 'active' doesn't necessarily mean that they are due to erupt imminently. The reason for the press coverage is perhaps understandable, because despite no 'supervolcanic' eruptions for over 26,500 years, there is a large *potential* for huge destruction. Supervolcanoes are sometimes known as mega-calderas because when they happen to have a big eruption, they efficiently empty all the magma stored in their chamber sitting below the ground, leaving behind a depression that the ground above collapses into, forming a caldera, the large hollow depression that forms in the ground after a major eruption. These types of volcanoes are classified by their size and magnitude, having had an eruption in their history that measures VEI 8 or greater, meaning they erupt over 1,000 cubic kilometres (240 cubic miles) of deposits.

While supervolcanoes are marked by their voluminous eruptions, they can also have smaller eruptions as part of their normal activity. These regions, when active, are also marked by geysers and hot springs, which is all part of the process of cooling off. The largest eruption at Yellowstone occurred around 2.1 million years ago, with a volume of 2,450 cubic kilometres (588 cubic miles). To put this into context, it is around 6,000 times larger than the most explosive of the Mount St. Helens eruptions, occurring on 18 May 1980. While this is very large, Yellowstone has had slightly smaller, yet still 'super' eruptions around 1.2 million and 640,000 years ago. However, these large eruptions are not the only ones Yellowstone produces. Around 70,000 years ago Yellowstone had an eruption at Pitchstone Plateau, a smaller event that allowed the magma chamber to let off some steam without a very big bang.

Nevertheless, it is the frequency of the large eruptions that gets people concerned. The media, in particular, like to play up to the idea that a supereruption at Yellowstone is 'due' anytime soon. This isn't accurate. In fact, Yellowstone might never erupt again. The problem is, though, that if it were to erupt another VEI 8, then it would have devastating effects locally and likely even globally.

One of Earth's more recent supereruptions, and one that was indeed very large, was that of Toba, around 74,000 years ago. Today, the site is marked by the beautiful caldera of Lake Toba, measuring some 100 kilometres (60 miles) by 30 kilometres (18 miles) and forming a part of the island of Sumatra in Indonesia. Following a gargantuan eruption – although it is known to have been preceded by at least three other large ones – there was a global winter lasting between six and ten years, and scientists suggest there is every possibility that humans themselves were nearly wiped out because of it. Evidence from the human genome has shown a rather restricted biological pool a few tens of thousands of years ago, potentially coinciding with this long phase of winter. It is understandable that we would have struggled to harvest enough nutritious food during this time if we were experiencing cold temperatures for many years. But did this mean that humans were reduced to just a few small groups who survived this period? For now, this is unknown, and fossil evidence suggests otherwise. Nevertheless, it's not an event I'd want to see in my lifetime, particularly as most of us living on the planet today survive in a much less sustainable way, requiring food to be transported around the world for our survival.

On the whole, the most problematic product of supervolcanic events seems to be ash, with the geological record showing that it has been common and troublesome throughout Earth's history. The Earth has also experienced supervolcanic eruptions that produce huge volumes of basalt. As we've seen, basalt tends to be slightly less hazardous than ash clouds and pyroclastic flows, yet in supervolcanic eruptions involving basalt, there is rather a lot of the stuff and it is erupted in geologically short timescales, around tens of millions of years or less, meaning these events can still produce enough lava to re-surface up to 1 per cent of Earth during that time. They are, quite simply, making new land in the fastest way we know possible.

Historically, these large basaltic eruptions have been known as flood basalts because, when viewed in the geological record, it would appear that the lava must have haemorrhaged from its source in very short timescales to produce such extensive and thick successions of now solid lava flows. We now think the term flood basalt is a little outdated, although it is often still used, because these eruptions are not really characterised by one fast-flowing flood of lava but rather a repeated series of large flows. As we saw in the previous chapter, today they are referred to as Large Igneous Provinces (LIP), a term that is still being refined; they are related to the arrival of a mantle plume below the crust. LIPs often begin life as a series of smaller flows that gradually build up to a main phase of activity consisting of repeated large flows of large volume and expanse from spatially restricted vents and fissures before gradually waning.

The largest examples defined by the Large Igneous Provinces Commission are given an 'A' rating if they erupted more than 100,000 cubic kilometres (24,000 cubic miles) over a short timescale of less than 50 million years. They are thought to begin erupting as low-volume flows that progress into the main phase of volcanism marked by the repeated eruption of large-volume, expansive flows before winding down.

One of the best modern examples to use as an illustration of this type of volcanism is the volcanoes that make up the Hawaiian Islands, or more specifically the Hawaiian-Emperor chain of seamounts. However, despite the Hawaiian hotspot being the most productive on Earth at the present day based on eruption rates, it is not classed as a true LIP according to the modern definition because it doesn't compare with the larger volumes of lava erupted by the historical LIPs. Nevertheless, the style of volcanism is thought to be very similar to that of LIPs and allows scientists to observe examples of the type of eruptions that occur and the edifices, such as shield volcanoes, that can form.

In particular, the Hawaiian hotspot is marked by sporadic, yet frequent, events of higher productivity. At the time of writing, the most recent of these was the lower Puna eruption in 2018, where long crustal fissures opened and produced extensive amounts of lava, yet still only travelling at speeds that you could, for the most part, outwalk with ease. Having said that, some of the Hawaiian lava did erupt as fast-flowing rivers of molten rock, so the term 'flood basalt', while not wholly accurate, is perhaps not far off in some cases.

Kilauea is one of the most active – and most visited – volcanoes on Earth, seen by three million people every year. It is easy to access. I was lucky enough to visit its active lava lake just weeks before the 2018 eruption began and I got to see a large volume of lava bubbling within the lava lake just before it completely emptied out onto the surrounding landscape via the fissures. To see molten rock bubbling and steaming away at the surface of the Earth was a sight I will never forget and, despite having studied volcanoes for years, this was the first and, so far, only time I've seen liquid rock. Yet, as we've seen, Kilauea is nowhere near as productive as LIPs. In the case of the LIPs, the fissures produced lava for sustained periods of years or even decades. This allowed them to erupt over 10,000 cubic kilometres (2,400 cubic miles) of lava, giving them a magnitude over 9 on the VEI scale, and covering thousands, if not millions, of square kilometres.

One of the reasons scientists have been concerned with studying LIPs, despite their basaltic lava being at the 'safer' end of the explosive scale, is for their potential to alter our atmosphere. Earlier we mentioned the 1783–4 eruption of Laki in Iceland. This gives us a taste of what we could expect from a LIP eruption at the present day, even though the Iceland plume is not defined as a current day LIP because it is not productive enough. A decade-long eruption of this type would be expected to cool the planet even more than the 1.3°C (2.3°F) decrease experienced by Laki. Although it might only take 50 years for the planet to recover from a nosedive in temperatures of anything up to 5°C (9°F), there would be immediate geopolitical and financial effects. But while the short-term effect is the cooling of our atmosphere, the long-term effect is warming. It's like climate whiplash. The effects of warming can be worse as it is harder for Earth to remove the problem. LIP eruptions have been associated with large mass extinctions on our planet in the past but, intriguingly, not all of them have killed off species and it is still not understood why the environmental effects are worse in some cases than others. Either way, the problem is usually related to warming ocean waters, which not only produces an environment unsuitable to many species, but also causes sea levels to rise.

Another reason we might want to study examples such as the Hawaiian Islands, or older examples of true LIPs on Earth, is that they are not unique to our planet and, in fact, our planet doesn't even have the 'biggest and best' examples, even if they are the most well known. Large regions of basaltic flows have now been recognised on other celestial bodies, including Mercury, Venus, Mars and the Moon. This feature is not confined to our own planet and that is because they are unrelated to plate tectonics. Such flows have played an important role in the geological history of the other planets too and constitute the most dominant form of volcanism in many of these places.

We will come on to the magmatic source of LIPs in the following chapter, where we will look at the ways in which these volcanoes construct our world, but while we know a lot about how we think they formed, there are still many uncertainties about them. The problem in understanding how LIPs are produced is that it is hard to know how frequent and extensive they've been throughout Earth history. This is, in part, thanks to plate tectonics and erosion, which has destroyed the evidence for some of our geological history. In this respect, and as we will see throughout this book, sometimes it is helpful to look at the surfaces of other planetary bodies where plate tectonics has not occurred, if we are to better understand our own.

The large volumes of lava that are present on the surfaces of the other inner planets, including the Moon, can, in terms of volume, be classed as LIPs. Such volcanism is thought to have played a significant, even main, role in the cooling histories of these objects. Fortunately, scientists can study these flows from afar, which is a good thing as they are extensive and best viewed at a distance in order to appreciate their expanse. They are often better preserved than their Earthly equivalents too, thanks to a lack of erosion and crustal recycling. The beauty of studying these planetary bodies is that we can remove the confusion of plate tectonics, which seems to be broadly unrelated to the production of LIPs. Nevertheless, the problem is that when we look at aerial images of their surfaces, while we can see that the flows are extensive, we can't necessarily tell how deep they are. In turn, this makes it tricky to work out the age of the flows, in absolute terms, and how long it took for them to be formed. Until that is possible, it will be hard to know if they were erupted in relatively short timescales of around one million years or so like our own LIPs, or whether these voluminous flows occurred over longer timescales.

While I've focused on some of the more destructive aspects of volcanoes in this chapter, this isn't the whole story. Volcanoes certainly have some wicked behaviours, but without them, the chances are that we wouldn't be here. Volcanoes are responsible for literally constructing our planet and providing and maintaining suitable conditions for life to survive on it, and possibly elsewhere in space. Because of this, in Chapter 4, we will investigate the more virtuous side of volcanoes and the ways in which they construct the environment we live in.

# CHAPTER THREE

# Making a Magma

To understand volcanoes on Earth, or even far beyond, we must comprehend how magma is made within a planetary object and how it finds its way to the surface. On Earth, and the other rocky objects generally found within the inner Solar System, magma is molten or semi-molten rock beneath the surface, which becomes lava when it erupts. The main technical difference between magma and lava is that magma is made up of melt (i.e. liquid rock), plus some solid crystals of common rock-forming minerals, along with bubbles of various gases, whereas in lava, although it is obviously similar since it is formed of the same material, the dissolved gases have mostly been lost on eruption, escaping out of the melt as the pressure is released when it gets to the surface and reaches lower atmospheric pressures.

This definition, however, need not apply to just those planetary objects made of rock, but equally to more icy worlds. As we've seen, the magma in these places might not be made of rock, but rather a less viscous version of whatever the body is made of, whether that be water or ammonia or something else; it is still technically classed as magma. The main differences we'll see on other planets and moons is that their surface pressure is different to that of our natural laboratory on Earth, sometimes very different in fact, which affects what happens to the magma and its dissolved gases when it reaches the surface. If lava is produced at the surface then this could mean that an eruption is more likely to ooze onto the surface, like toothpaste from a tube, or pop, rather than explode with ferocious power.

The reason we have a great deal of basaltic rock on Earth is because this is what is formed when the interior of the Earth, the mantle, melts. So, basalt becomes our most common type of lava. But the Earth's innards are not themselves made from basalt, so first we need to investigate what is inside the Earth in more detail before we can understand why it doesn't contain basalt throughout.

The **mantle** is the middle layer of the Earth, which formed when our young and largely molten planet separated itself into a core, mantle and crust. The **core** was formed as the heavier elements, mostly iron, sunk to Earth's centre. The **crust** then formed initially as mostly basaltic in composition, as molten magma ascended from the parts of the mantle that melted at lower temperatures and reached the surface, leaving minerals with higher melting points behind. Through repeated recycling within the crust, easier to melt minerals started to accumulate in the thickest parts, which became the continents, changing the chemistry of the continental crust over time, while the thinner oceanic crust is still basaltic.

As we saw in the previous chapter, the mantle constitutes a large proportion of the volume of our planet and it is thanks to its internal flow that all of the geological surface activities are possible. If we were to crack open the Earth, it would be mostly green inside, until we reached the core. Well, once it had cooled down, that is. If it was still hot then it would glow, but after cooling we would see that it is composed of a green mineral called olivine. Along with the olivine there is another mineral called pyroxene, which helps to make up the bulk of the upper mantle, forming a rock type known as peridotite.

Peridotite has a lower silicon dioxide, or $SiO_2$ (otherwise known as silica) content and is often described as 'ultramafic',[*] whereas the slightly higher silica content of basalt makes it 'mafic'.[†] We know that the mantle is not homogeneous (it is not the exact same composition throughout), partly because the Earth constantly removes parts of its insides around to make volcanoes and it also adds material back in at subduction zones. The mantle is a busy place, and all this periodic removal and addition of materials means that it is not very well mixed because it convects rather slowly and inefficiently. Any heterogeneities that are created from melt removal or addition of lithosphere at subduction zones are mixed in rather slowly, even on long geological timescales. Nevertheless, to represent the average, or bulk, composition of the mantle, scientists use a rock of peridotite composition as this tends to represent those parts of the mantle that melt and make their way to the surface. The other terrestrial planets – the rocky ones within the inner Solar System – also have mantles that have melted in the past and produced basaltic magmas. Basaltic magma is, therefore, extremely common throughout the inner Solar System.

## Melting a planet

Producing a liquid from the peridotite mantle – that is, melting it – is not as easy as you might think. Despite the centre of the Earth being hotter than the surface of the Sun,

---

[*] Igneous rock with very low silica and potassium content, but with very high magnesium and iron. The Earth's mantle is composed of ultramafic rocks.

[†] Igneous rock that is rich in the heavier elements such as magnesium and iron, but not as rich as ultramafic rocks. The term is derived from MA (magnesium) and FIC (from the Latin for iron). These rocks are often produced by the melting of the mantle to form gabbro or basalt.

at around 5,000°C (9,032°F), the temperatures within the upper mantle are not actually hot enough for peridotite to melt. It's not as if the whole Earth is just a big reservoir of molten rock impatiently waiting for a crack to appear in the crust for it to spew out from. The Earth experiences a natural increase in temperature from the surface to its centre, which is known as its geothermal gradient. In general, and away from the plate boundaries where crust is made or destroyed, Earth's increase in temperature with depth is, on average, on the order of 25°C per kilometre. Nevertheless, this number only really applies to the upper 100 kilometres (60 miles) of the planet, the rigid lithospheric portion, and we must remember that the centre of our planet is over 6,000 kilometres (3,700 miles) from the surface. After we get through the lithosphere, the temperature increase with depth actually drops off through the mantle, before increasing again at the base of the mantle and towards the core. We must also bear in mind that there are quite large global variations in geothermal gradient dependent on the tectonic setting. For example, if you are burrowing down into the Earth near to a spreading ridge, where magma is coming to the surface, then you will find that the temperature increase with depth is a lot more rapid than it would be in the middle of an ocean basin far from a plate boundary, otherwise known as an intraplate setting.

Equally, different planets are marked by wildly different thermal gradients, which is the result of the history they experienced since they formed out of the interstellar ether 4.5 billion years ago. We'll cover this in much more detail in Chapters 7 and 8, as the way a planet heats up and cools down has important consequences for its evolution, including volcanic activity. Interestingly, scientists don't know exactly how much heat is stored within each planetary object at the present time. From afar, it is possible to study how a planetary object radiates heat into space, and by comparing this to calculations on the amount of heat received from the Sun, it can give a reasonable estimate of how much heat it produces from its interior. For example, at the present day, Uranus, one of the furthest planets from the Sun and sitting in the cold outer reaches of the Solar System, only radiates out as much heat as it receives from the Sun. From this we can conclude, rather simply, that it has cooled down and no longer hosts an internal heat source.

But in order to make an accurate estimate of a planet's heat flow we need to be able to land a spacecraft and drill down to place a thermometer below the surface. Even for Mars, a planet that we have visited with numerous spacecraft over the years, it was not until 2019 that we began to make our first attempts to measure its temperature, when the NASA InSight mission that landed in 2018 started to drill down into its outer layers. At the time of writing, the 'mole', which was the instrument burrowing into Mars, encountered some problems related to the fact that the Martian rocks were 'stickier' than expected. This was a term the scientists used to account for the fact that their attempts to drill to the required depth for Mars' temperature to be taken with accuracy were hindered by the rocks they encountered. But it just shows that even with the best planning over many years of research, studying other planets remotely, or with robotic spacecraft, is fraught with challenges.

Matters were slightly different, and easier, on the Moon because we had the Apollo astronauts to help out. Astronauts on Apollo 15 and 17 placed four temperature-sensing probes between 1.6 and 2.3 metres (5.2 and 7.5 feet) deep in the Moon's crust as part of the 'Heat Flow Experiment' that ran from 1971 to 1977. After the data were analysed, scientists found that the Moon's heat flux was around 20 per cent of Earth's average. This lower heat flux was not surprising, since the Moon is no longer an active body, plus it is a lot smaller than the Earth and is therefore expected to have cooled quicker (more detail on this in Chapter 7). But the results indicate the Moon is not completely frozen, which comes as a slight surprise because its surface appears completely inactive. Not only does the amount of heat flowing out of a planet tell us about the state of its interior, it also tells us something about its overall composition. The Moon's heatflow, in fact, suggests it might retain some molten material inside.

Despite the majority of the interior of our planet being solid, it is possible for portions of it to melt, but it is not always caused by something as straightforward as increasing its temperature. The point at which peridotite melts is called the solidus. If we were to take a lump of peridotite and try to melt it at the surface of the planet, we would need to heat it until we reach its solidus temperature, which is around 1,300°C (2,372°F), when it starts to turn to liquid. However, if we travel deep into the Earth, we find that the increase in pressure acts to inhibit this melting. The result is that the deeper the peridotite, and therefore the higher the pressure, the higher the temperature we would need to melt it. But clearly, we get liquid rock bubbling up from the Earth's interior as lava at the surface, so something must help it transform from solid to liquid. It turns out to not all be about temperature.

## Salting the ice

There are three main ways to melt the inside of a planet. One way involves adding heat, as discussed above; another involves releasing a bit of pressure; and yet another involves adding a substance that acts to artificially lower the melting point, like adding salt to water. Taking the last first, if we want to make peridotite melt, even at depths and temperatures where it should still be solid in the Earth's mantle, we can add a substance such as water. We can think of this like adding salt to ice because it has essentially the same effect: we are adding an impurity that changes the chemical composition. For peridotite, water acts as ice's salt by artificially lowering its solidus. This means that melting can happen at temperatures where the rock would normally be solid. The main place we add water to the mantle is at subduction zones, where portions of the oceanic lithosphere are transported down until the water they contain is forced out of them, infiltrating the mantle above and causing it to melt. This is why volcanoes then appear above, and adjacent to, subduction zones, as those freshly made, buoyant mantle liquids make their way up to the surface and seep through the crust above. We'll learn more about this process in the following chapter. But we don't always need water for this effect to take place, as other substances such as carbon dioxide can have the same effect.

Releasing the pressure on the mantle allows it to decompress, which is the second way that melting can happen. The way this can be achieved is when a portion of peridotite naturally rises in the mantle so that it finds itself at a lower pressure. If this happens slowly then the mantle will cool on its way up, counteracting the effects of the lowering pressure, and it will not melt. But if it rises rapidly, it can remain at, or almost at, the same temperature it was when at depth. The classic example of this is a mid-ocean ridge where the lithosphere – the lid holding the pressure in – is pulled apart because of plate tectonic motions. The space that is left behind between the plates is filled with mantle that upwells from below because it moves to fill the void, suddenly experiencing a pressure drop. However, not yet having lost its heat, it is able to cross its solidus and ooze out at the ridge as liquid rock.

Of course, the third way to melt the mantle is to heat it, increasing its local geothermal gradient. But to do this we need a source of heat and it needs to be transferred from one area to another, which is usually by conduction. One place this could happen is at the very base of the mantle where the heat from the core warms the mantle sitting above it. In some cases, scientists think this can produce enough heating to make a significant portion of lower mantle rock cross its solidus and buoyantly rise. It will do so because liquid rock is less dense than solid rock. If this happened, we would call it a mantle plume, a chimney of hot liquid rock that crosses from the deep to shallow Earth. Mantle plumes can account for the hotspot volcanoes that appear at the surface, such as in the Hawaiian Islands. Because the mantle plume rises relatively quickly from the deep, and brings heat with it, its warmth is conducted away quite easily in the shallower depths of the mantle when it stalls because it has nowhere to go, trapped in by the lithosphere. This is where the difference in temperature between it and the surrounding material is largest, which in turn causes melting of the surrounding peridotite. Mantle plumes are a fairly common intraplate feature on Earth and an inherently common feature of the other terrestrial planets: Mercury, Venus, Mars and even Io.

Wherever we are in the Solar System, and whatever a planetary body is made of – ice or rock – in order to produce liquid at their surfaces they must use at least one of these three methods to generate melting of their interiors. Of course, the solidus temperatures and pressures for each planetary body are different depending on what exactly

they are made from and what state it is in (since some
regions of the interior of a planet might already be liquid,
such as Earth's outer core). If we take Enceladus, one of
Saturn's moons, as an example, then while it is thought to
have a rocky core, it might also possess a subsurface liquid
water ocean capped by a relatively thick crust of solid ice.
So we must be prepared to expect the unexpected when
we look at other planetary bodies.

## Forming liquid

Whatever the mechanism for melting the mantle, in reality,
we find that it melts rather gradually and does not all go
from solid to liquid in the blink of an eye. In fact, most of
the igneous rocks we see at the surface are the result of
partial melting, that is, only some parts of the solid mantle
melt at one time. The reason for this is that the mantle is
formed from a range of minerals, each with their own
unique melting temperature. Peridotite, for example, is
composed of mainly two minerals, olivine and pyroxene,
that each have their own melting conditions. Nevertheless,
although magmas can vary wildly in composition, they all
have the same main elements: oxygen, silicon, aluminium,
iron, calcium, sodium, magnesium and potassium, that
combine in different proportions to form a range of
minerals. When the rock melts, its course of melting will
depend on what exactly it contains.

The first parts of the mantle to melt are, unsurprisingly,
those portions made up of rock minerals with the lowest
melting points. Once these begin to melt then small
amounts of liquid rock are produced that get trapped
between the unmelted, still solid grains that have a different
composition and higher melting point. The melts that are
made first start to be squeezed out slowly from between the

grains and move along their boundaries. One useful feature of a melt is that it tends to be around 10 per cent less dense than the rock it came from, making it more buoyant and so allowing it to rise.

The first magmas produced from the melting of peridotite mantle are usually broadly basaltic in composition. As we briefly mentioned above, when different igneous rock types are discussed, geologists tend to express them in terms of their silicon dioxide, or silica, content. One of the reasons is that silicon and oxygen are some of the most common elements in the Earth and they easily combine to produce silicate compounds based around the silicon tetrahedron structure $(SiO_4)$, a silicon atom surrounded by, and bonded to, four oxygen atoms that define the corners of the tetrahedron shape (it looks like a triangular pyramid). Silicates can also incorporate other elements into their crystal mix, such as magnesium and iron. In doing this, a whole host of different mineral compositions can be generated such that silicate minerals constitute over 90 per cent of the Earth's crust.

It is the added elements in silicates that change their chemistry and structure, and therefore their melting point and other chemical properties. For simplicity, the silica content given for a rock type assumes that all the silicate minerals were simple compounds of just silicon and oxygen, and it ignores the other elements that might be involved. For the mantle, this gives a silica content of around 45 per cent, making it ultramafic. One of the most important features of silica in molten rocks is that it acts to bind them together, which also has implications for the way the lava erupts. The more silica in the rock, the more viscous and stickier it will be. At very high temperatures most magma is liquid simply because there is so much energy in the system that the atoms can't bond together.

This is usually the case for temperatures over 1,300°C (2,372°F). But as the temperature starts to come down, perhaps because the magma is moving, the silicon and oxygen combine to form silica tetrahedra. The key part comes about as the magma further cools because the tetrahedra start to link together to form chains. This is known as polymerisation and has the important effect of making the magma more viscous.

When basaltic lavas are produced from the mantle, they have a slightly higher silica content than the mantle they leave behind, usually giving them a silica value of around 49 per cent, which makes them mafic. The reason for this is that, in reality, it is impossible to melt the entire mantle source completely. The residue, or solid part, that is left behind after melting begins has a lower silica content than the early melt. The relatively low silica content of basaltic magmas is an important feature as it makes them some of the most 'fluid' lavas the Earth can produce. But despite their low viscosity, they are still 100,000 times more viscous than water. Their consistency could be compared to slightly lumpy porridge, so you can see why you might be able to outrun them. The other property that makes basaltic lavas less dangerous is that they're not very gassy and are, therefore, less explosive.

Basaltic rock is not just common throughout the inner Solar System, but many asteroids are basaltic in composition too. The reason is that most of these planetary bodies were formed in the same part of the Solar System, relatively close to the Sun, being built of the same basic building blocks as each other at around the same time. Therefore, when these planetary bodies are melted, the magmas they produce are similar in composition. As we've already discussed, what differs is the conditions these magmas meet when they reach the surface of their planetary body during eruption.

The environment into which they are extruded will control whether the lavas flow quickly or not, and also whether they will erupt explosively.

But basaltic lavas are not, in fact, the most fluid lavas found on Earth. Although relatively rare, there are some rather unusual lavas that are even less viscous than basalt. Carbonatite lavas are rich in carbon and carbon dioxide and often packed with sodium and potassium, but with a very low silica content, usually around 10 per cent. Hence, they lack that important structure provided by the silicon tetrahedra and are, therefore, very fluid.

Carbonatite lavas are much cooler than their basaltic buddies and sometimes they are classed as 'cold', in relative terms anyway. There are tales of people having fallen into a carbonatite flow and survived, but as they are around 500°C (900°F), I'd personally still be wary of them. There are only a few places on Earth that have erupted these lavas, and they are confined to continental rift settings – where two continental lithospheric plates are being pulled apart from one another – such as in East Africa where the two tectonic plates, the Nubian and Somali plates, are diverging at 6–7 millimetres (0.24–0.28 inches) per year. This is the same type of tectonic setting as a mid-ocean ridge except that it is on land. Given some time, the continental rift here will progress to become a mid-ocean ridge as the plates continue to part and an ocean floods in to fill the gap where new basaltic crust is being made. One of the most famous carbonatite volcanoes is Ol Doinyo Lengai in Tanzania, which is active at the present day as part of the Gregory Rift, one of the smaller rifts that helps to make up the volcanic system of the East African Rift. It is quite a sight too, as while these lavas appear dark when molten and erupting, they cool to a stark white colour, much like a limestone, thanks to the high proportion of carbon.

Carbonatite lavas haven't been found elsewhere in the Solar System for certain … yet. This is, perhaps, unsurprising because we think that Earth is the only planet to have continental crust, which is a later generation of crust to form on a planet, and for which it is believed plate tectonics is necessary. We'll come back to, and explore, this idea in future chapters. Nevertheless, the fact that scientists have observed these rather strange and very fluid carbon-rich lavas erupting on Earth means that they've been able to start looking at other planetary surfaces and wondering whether the volcanic features they see might be related.

For example, at first sight Venus appears to have long, sinuating rivers crossing its surface in some places, but we know the planet is far too hot to host liquid water. Instead, scientists think that these river-like features are formed by lava flows, but lava that must have been very fluid to behave in such a way. The question then remains: are these lavas a different composition to the standard basalt we expect, or are they of the same composition, but capable of flowing further, with seemingly much lower viscosity, because the surface conditions are so different? We know that the surface pressure of Venus is crushingly high, and its temperature is enough to melt lead, which suggests that basaltic lavas could certainly be expected to behave differently on eruption from how they would on Earth. However, although this is still unproven without further space missions, some scientists have suggested that carbonatite magmas might account for some of the special volcanic features observed on Venus. Unfortunately, the surface of Venus remains a bit of a mystery for now because it is an incredibly challenging environment to send a spacecraft – the Russian Venera Venus landers lasted just a few hours, at most, on the surface before being crushed and rendered useless. Without a mission that can sample the flows, or a meteorite arriving on Earth that has originated from the correct region of Venus, its long and fluid lava flows will remain a mystery for now.

## Pools of magma

When we hear about a volcano, we often also hear about the magma chamber that sits beneath. Most people might have the idea in their head of a large pool of liquid rock bubbling away just metres or kilometres below the volcano, which fuels the eruptions seen at the surface. This too was essentially the way that scientists viewed magma chambers for many years, perhaps with a little more nuance, but the idea was similar. However, recently there has been a bit of a shift away from this concept. Scientists suggest that large magma chambers might not even exist at all. Instead, as scientists have completed more and more detailed geophysical surveys of the ground beneath and surrounding individual volcanoes – allowed partly by the improvements in technology in recent decades – they have discovered that the crust is made up of a series of lenses of magma, bodies of molten rock that are broadly lens-shaped and measure from metres across right up to kilometres, that are all interconnected in a complex plumbing system. It's even suggested that a lot of this magma isn't liquid, but a mix of molten rock, or melt, and a large volume of crystals that have solidified out of the melt. If this melt-crystal mush is more than 50 per cent solid crystals, then it's thought that the volcano is unlikely to be capable of erupting because the crystals lock together, freezing the system in place and preventing movement of the flow of magma to the surface.

The magmatic plumbing system that feeds a volcano at the surface can be very complicated – almost as far from the idea of a simple, large pool of liquid as we could get – including offshoots, pipes and dead-end lenses. And as we've seen, there's also the possibility for solid, liquid and crystal mush mixtures in different places and at different times, which means that forecasting when and where magma will be released, and in turn if that will produce a large or small eruption, is not simple. As we combine years of data collection on the physical and chemical changes that occur during, and leading up to, volcanic unrest, scientists are finding that their forecasting of eruptions on Earth is becoming more accurate. However, when it comes to other planets, unless we have been able to observe a periodicity to eruptions, as has been possible on Io (which is especially active), then forecasting when and where volcanic activity is going to take place is anyone's guess.

One of the problems is that not all of our Earth studies translate across to space because the environments are so different, and we certainly don't have the geophysical data to piece together the potentially intricate plumbing systems out there, since we haven't yet been able to do that for even half of the volcanoes on Earth. The best we can do is to look back at the geological record of these planets to see when lavas have flowed in the past, and if they appear to have done so on a regular timeframe. But we can also observe planetary bodies to look for signs of current unrest, which, as we saw above, has only been possible in one place, Jupiter's moon Io. Our apparent luck at spotting volcanic eruptions at Io is mostly related both to the fact that two missions have performed fly-bys and because it is incredibly active. If we want to capture eruptions occurring on other planetary surfaces and obtain a much deeper understanding of their volcanic future potential, then there is only so much we can achieve with a fly-by; if we are to delve into their geological records then we really need rock samples. Nevertheless, a great deal can be achieved with orbiters and landers.

## Releasing the pressure: stickiness

I've said that when we melt the Earth's mantle, or the mantles of other terrestrial planets, we generally produce basalt, so you may be wondering how we get more silica-rich, 'sticky' lavas. After all, the volcanoes hosting these sticky lavas are the ones that generally cause us the most trouble, because stickier lavas tend to be more explosive. These are generally the 'classic' steep-sided volcanoes which, although they erupt less frequently, do so with much more force. They are often known as composite, or strato, volcanoes because their slopes are composed of interspersed layers of ash and lava.

A magma becomes more silica-rich, which can fuel explosive eruptions, depending on what exactly was melted to produce it. If a piece of crustal rock is melted instead of standard mantle, then a much more silica-rich magma is produced. The reason for this is that crustal rocks themselves have higher silica contents than the mantle because they were formed by melting of the mantle, and as we learnt earlier, a melt is always more silica-rich than the rock it leaves behind, its parent rock. Nonetheless, even if we begin with a basaltic melt that came from the mantle, there are ways that it can be transformed into an even more silica-rich, and thus stickier, lava. The geological processes that help to make lavas more silica-rich occur as the magma journeys to the surface. These processes have aided in the generation of an array of weird and wonderful, and sometimes very explosive, volcanoes. They are also responsible for forming the continental crust, which is of utmost importance to our existence.

As we saw, a magma will tend to rise through the crust because it is less dense than the surrounding rock and is, therefore, more buoyant. So, a basaltic magma will want to rise away from its source. Nevertheless, this isn't an easy process. During a magma's somewhat troubled movement through the Earth's crust, it can become stalled at various points, sometimes for millennia. Whenever this happens, the magma tends to sit around and starts to fester, or 'evolve', for a more scientifically accurate term. What this means is that the magma changes in composition. It might mix with other batches of magma – of the same, or a different, composition – that are also rising through the crust to meet it. It might pluck pieces of rocks from the Earth's subsurface along the way and incorporate them by melting.

It might also start to crystallise minerals as it cools down during its rise, which is inevitable because the surrounding crustal rocks are usually cooler. When a magma cools slowly, crystals of various rock minerals are produced in a process that is the same as melting, but in reverse. The minerals with the highest melting point are the first to crystallise. In fact, sometimes a magma never reaches the surface but stops, cools, crystallises and solidifies in the crust before it can erupt. If the original magma was silica-rich, on the way to fuel an explosive volcanic eruption before it got stopped in its tracks, then the well-known rock it produces within the crust is granite.

The crystallisation of a magma follows a predictable progression that has been studied in detail by scientists for years as it allows them to predict the minerals that are likely to occur together in a rock. This means scientists can figure out how, and where, a rock formed. This process is commonly called Bowen's reaction series after experiments carried out by the petrologist Norman L. Bowen in the early 1900s. His experiments formed the basis for scientists to be able to study a rock, see what minerals were present, and then calculate the temperatures and pressures at which the rock crystallised.

As crystallisation progresses, the different minerals within a cooling magma solidify, and in doing so they remove silicon dioxide, along with magnesium, iron and other elements that are required to form the crystallising mineral. The important point is that the first minerals to form have lower silica than the melt from which they originated. The melt left behind is, therefore, altered from its original composition, becoming more silica-rich.

Sometimes these newly formed crystals settle out of the magma by sinking through the liquid because they are denser. This can only happen, however, if the liquid is not too viscous, otherwise its stickiness creates a physical barrier for the solid crystals to be pulled down by gravity. However, if and when the crystals do settle out, they leave behind a more silica-rich melt. If this process continues as the magma cools further, then the liquid becomes progressively more silica-rich, viscous and sticky. It also makes the magmas gassier, which we'll come on to shortly as this has important implications for the way the lava is subsequently erupted.

As magmas become more silica-rich, it is possible to produce compositions such as andesite and rhyolite. Both are packed with the structure-forming silica bonds that make the lavas more viscous and, therefore, harder to move. Andesite ranges from 52 to 63 per cent silica, and rhyolite over 68 per cent. All those silicon and oxygen bonds literally bind these magmas up, making them hundreds to millions of times more viscous than a basaltic magma. The result is, if andesitic and rhyolitic magmas manage to work their way to the surface to be erupted out of a volcano, then they are too thick to flow as a basaltic lava might, tending to ooze out much more slowly. Nevertheless, this doesn't necessarily make them less dangerous.

## Releasing the pressure

We know now that increasing the silica content of a lava has far-reaching consequences on the style of eruptions that occur, making them much more explosive and dangerous. However, there is another component of magmas that also plays a key role in their potential for explosive behaviour. All magmas contain gases known as volatiles. These can be in the form of water and carbon dioxide but also sulphur, chlorine or fluorine, to name just a few of the more common ones. Magmas that form from melting of the mantle tend to have lower volatile contents than those formed by melting of crustal rocks, simply because the mantle contains fewer volatiles to start with. So, the way in which the magma is produced controls how sticky, gassy and explosive it will be. You make your lava less sticky by adding volatiles, but then it becomes potentially more dangerous because those volatiles become gases that create explosions at the surface. Understanding the abundance of volatiles in a magma is important as they govern the eruptive behaviour of the whole volcano along with its temperature, pressure and general composition.

When volatiles are dissolved within a magma, they don't really pose a problem. As a volatile-laden magma rises through the crust, because it is buoyant, the pressure acting on it decreases and these volatiles exsolve, or escape, from the liquid to form gas bubbles. Basically, as the confining pressure decreases, the bubbles grow and expand, which can create a magmatic froth. The bubbles don't have anywhere to go if the magma is still trapped within the Earth's crust. They might move up to the top of the 'magma body', whether this is a chamber, lens, crack or the conduit (the pipe connecting the magma body to the volcanic vent on the surface). However, the bubbles can only migrate through the magma if it isn't too viscous, otherwise they will remain trapped in the liquid. Carbon dioxide tends to exsolve at a deeper level, so it usually has more time to rise to the top of the magma body and then seep out before eruption. Water, on the other hand, stays dissolved until shallower depths and thus tends to be the main eruptive volatile.

The problem with gases remaining in the liquid is that they pose a greater risk. A more viscous lava, such as andesite or rhyolite, can build up more gas because the bubbles face more resistance escaping; the gas is literally stuck within the molten rock. At some point, as the volatiles continue to exsolve, the pressure within the magma increases until the bubbles cannot be contained any longer. As the bubbles attempt to burst out of the magma, or the chamber in which they are held, and travel to the Earth's surface, they continue to expand. As the pressure drops even further on the way up, their rapid expansion can cause an explosive eruption. We can think of this set-up a little like shaking up a champagne bottle. No bubbles are visible as long as the cork is held securely in place by the wire, but once the cork is popped off, the excited bubbling liquid shoots out. While this is a bit of a waste of perfectly nice sparkling wine, it's certainly fun. However, the situation is less fun at a volcano because we are talking about scalding molten rock. Of course, once the gases are released from a magma, it instantly becomes more viscous again and, therefore, a bit less dangerous, a bit like our champagne, which we can now pour into glasses (what is left of it anyway).

When volatiles escape from a very evolved and silica-rich magma, its stickiness means that it is likely to be exploded into pieces. A lava can fragment into tiny pieces, forming small shards of incandescent rock that can be blown out of the volcano at several hundred metres per second. The small size of the fragmented pieces of lava means that they cool quickly as they travel back down to the Earth's surface, producing volcanic ash and pumice, which are essentially pieces of frothy exploded magma.

If we take a basaltic magma, which is usually less viscous, then we can understand how it allows gas bubbles to escape gradually such that basalt is less likely to produce explosive eruptions. Basaltic lava tends to exude in a gentle effusive flow. Nevertheless, explosive eruptions can be produced by basaltic magmas and when this happens, a fire fountain, or a spray of lava, is produced. These are impressive features that, while they are literally spraying molten lava out of the Earth, don't erupt with as much force as ashy eruptions and are, therefore, less dangerous.

We can think of fire fountains a little like a geyser of molten rock; they can be observed in many places, but Stromboli volcano in Italy and Kilauea on Hawai'i are well-known examples. While it is a spectacular sight, the rock isn't on fire, of course; it is simply 'red hot'. The blobs of lava that are shot out of the volcanic vent or fissure cool rapidly because of their large surface area to volume ratio, and can rain down as solid, yet still hot, rock.

Where we find interesting volcanic features on Earth, scientists hope that our cosmic neighbours might share similar characteristics. Sure enough, even if we just focus on our only satellite and closest neighbour, the Moon, it is thought to have erupted fire fountains in its past. During the early 1970s the Apollo astronauts collected tiny glass beads – flecks of lava less than a millimetre in diameter – from the lunar surface; specifically, during the Apollo 15 and 17 missions. These represent the products of lunar fire-fountain eruptions that are thought to have been fuelled by carbon monoxide as their main volatile. The glass beads formed because the lava cooled so quickly that it didn't have time to crystallise minerals and was instead quenched straight into a glass. When larger pieces cool more slowly, they have time to crystallise minerals and turn black, like normal lava. Of course, the Moon appears grey and dormant today, but these special glass beads reveal its more explosive and fierier past, even if that was 3.5 billion years ago.

Fire-fountain eruptions are also known as pyroclastic eruptions. The term 'pyroclastic' comes from the Greek for 'fire' and 'broken', so refers to the fragmentation of rock. All the explosive volcanic products discussed above are some type of pyroclastic deposit, including glass beads, ash and pumice. Sometimes larger lumps of intact magma get ejected from the volcano to form chunks of rock called volcanic bombs. Such material often comes under the umbrella term of ballistics, for obvious reasons. Volcanic bombs can form intriguing shapes, which are created as the still molten lava hurtles through the air and begins to cool. Some can adopt a spindle-like shape and others more rounded morphologies. Volcanic bombs don't always have time to cool completely before they impact the surface and, in these cases, they can form a sort of lava cowpat.

It is estimated that around 40 per cent of fatal incidents within the first 5 kilometres (3 miles) of a volcano are caused by ballistics. Because of the mass of these bombs, it is unlikely for them to be propelled further than 5 kilometres (3 miles) from the summit, but not totally unheard of. However, because of the low probability of large numbers of people being within the 5-kilometre (3-mile) zone, ballistics incidents account for less than 1 per cent of all volcano deaths worldwide. This is a stark reminder of why we need to be very careful the closer we get to the summit of a volcano, because that is where you are in most danger from being impacted by hot, molten or solid volcanic projectiles.

CHAPTER FOUR

# Construction

Fortunately for us Earthlings, the material products of volcanoes – namely lava and ash – bring many benefits, despite their potentially life-threatening drawbacks. Volcanoes are responsible for constructing new land when they erupt lava or ash, and these materials flow or rain down on the surrounding area. The result can be to cover over existing terrain, extend a coastline into the sea or even form a new island.

The island of Surtsey, just south of Iceland, is one of our planet's younger islands, having been born out of the sea in 1963. The eruption that produced Surtsey was captured on television. This meant that many people got to see this special event occurring, at a time when television broadcasts were relatively new and still exciting. The eruption started a few days prior to the first piece of new land being spotted, as it began 130 metres (425 feet) below the surface of the sea. There were a few warnings that something was happening though. Firstly, earthquakes were detected in the preceding weeks on the seabed in the location where Surtsey appeared. Secondly, people living in Vik, an Icelandic coastal town about 80 kilometres (50 miles) away from the newly forming Surtsey, reported that they could smell rotten eggs, indicative of hydrogen sulphide gas: a common volcanic gas often remarked upon because of its distinctive smell. In fact, even without Surtsey's eruptions, if you go to Iceland and you have a sensitive sense of smell (as I do), then you will smell hydrogen sulphide all over the place. Thirdly, fisherman working in the region found that the sea temperature had increased by a couple of degrees, which is a worrying amount when you are trying to catch fish that thrive in cold waters. As Surtsey initially began to grow underwater, any explosions related to the early eruptions on the seafloor were suppressed by the depth of the sea. An interesting point if we are considering eruptions on other planetary bodies, where the surface pressure is higher than on Earth. Gradually though, flow after flow erupted out onto the seafloor, covering over previously erupted material and building up a mound that soon reached the surface of the sea.

The first signs of Surtsey's appearance were noticed by fishermen on a trawler, who reported black smoke in the sea. They apparently assumed a boat was on fire. It was then noted that the black smoke was accompanied by explosions and columns of ash. People quickly realised that these events were marking the start of a new undersea eruption that was fast approaching the surface. The eruptions that created Surtsey were extremely violent because the lava that was being extruded initially below the ocean instantly met cold seawater, where it rapidly cooled and occasionally exploded into tiny pieces of loose rock. Such events are known as phreatomagmatic eruptions, where the molten rock explodes on contact with very cold water, such as is found around Iceland. These tiny pieces of rock are known as scoria – frothy pieces of basaltic lava – and they continued to build up the island above the surface of the water.

While such geological action has occurred all over our planet for millennia, having the option to view it at the present day offers some valuable opportunities. Such new volcanic islands can remain almost untouched by humans either because they are too unstable to be accessed, or because they are legally protected as new nature reserves.

They give botanists and other biologists the chance to study their flora, and sometimes fauna, as lifeforms gradually colonise the pristine land. It's hard to find parts of the Earth that are so fresh, unspoilt and untouched by human interference. You'd be surprised how quickly plants start growing on uninhabited land formed simply of lava and ash. It doesn't take long for a few windblown seeds to nestle down into cracks in the lava. In Hawaii there is a well-known tree that is the first to colonise a fresh lava flow. It is called 'Ōhi'a (pronounced oh-hee-ah) and is initially a lone voyager on an otherwise barren volcanic surface. However, with its bright red blossoms and green leaves it supports the appearance of further life on the lava flows, representing the cornerstone of a new ecosystem.

Islands that suddenly appear out of the sea can rapidly disappear back into the sea owing to rapid erosion at sea level. In fact, some scientists suggest that Surtsey could disappear below sea level within the next 100 years as eruptions have ceased but erosion continues. It is a fine balance between an island growing large enough to house organisms that help to protect it, binding its surface together, and one that doesn't grow quite high enough to avoid being consumed back into the sea. Kavachi volcano in the Solomon Islands of the south-west Pacific is one of the world's most active submarine volcanoes. After it was first recorded in 1939, it has emerged and been eroded back into the sea at least eight times, and this was despite the fact that its 2003 eruption produced an island 15 metres (50 feet) high.

Even if such young volcanic islands sometimes have a tendency to vanish almost as quickly as they appear, they still have their uses to scientists studying other planets. Mars is well known to have been volcanically active in its past and it's also thought to have hosted persistent bodies of water on its surface, meaning the red planet was once blue. If Mars' volcanoes had erupted into water as some have on Earth – and we have every reason to suspect they could have – then studying the way in which our own volcanoes have interacted with water can tell us about Mars' past environment. For example, as noted by the fishermen, the seawater around Surtsey was heated by the eruption of the undersea volcano. You might not think this is something we could spot in the geological record on another planet if we have no eyewitness to tell us it happened. But this is where the rocks themselves become our witness account. The slightly raised temperature of the seawater acts to speed up reactions between it and the minerals present in the newly erupted volcanic rock. This is actually a well-known process known as palagonitisation, and it creates a new, highly erosion-resistant material known as 'tuff'. Such complex chemical reactions literally help to make the island tougher (excuse the pun), preserving it for longer than would otherwise be expected.

This is where Mars comes in. Infrared spectroscopic studies have detected the presence of palagonite on its surface. This substance is found as a component of dust in much of the Martian regolith, the material like soil, but without the organic matter provided by (mainly plant) life on Earth, covering the surface of Mars. Thanks to studies of our own planet's infant volcanic islands, the presence of palagonite on Mars has been cited as evidence for the existence of water in the Martian past: water that was heated and reacted with the volcanic material that was building new volcanic islands.

At the present day, new islands still sometimes appear out of Earth's oceans, and so it is a process scientists can get to study in more detail each time one can be observed in real time. In 2014, the eruption of an undersea volcano in the South Pacific Kingdom of Tonga rose above the surface of the water. It is now called Hunga Tonga–Hunga Ha'apai. It is the first such island to form in the modern era of advanced satellite technology, so it has given scientists the opportunity to study it remotely, just as we might study another planet from orbit. The advantage is that these fresh volcanic islands need not be visited by humans at all, yet the processes that form and destroy them can still be monitored in detail from afar. They also give scientists the chance to observe how microbial life interacts with the new land that sits at the land–water margin. The beauty of Hunga Tonga–Hunga Ha'apai is that it allows scientists to study this process as it's happening, in real time, rather than relying on the geological record. As we've seen, Mars is thought to have had water flowing on its surface in the past and a history of active volcanoes, so being able to watch on Earth how those two things interact over time is important for our understanding of Mars' geological history.

## Moving apart

If we were running a contest for the most productive construction company on Earth and I were the judge, I would have to give the prize to volcanoes. Our planet is constantly constructing new crust and land. In another prize category, the winner of the most productive type of volcano would go to those on the spreading ridges. They are akin to a mega corporation of volcanoes, looking almost nothing like the classic conical, snow-capped mountains that often spring to mind when we visualise fiery volcanoes. As we've already learnt, a spreading ridge, such as that which forms mid-ocean ridges, occurs where two tectonic plates are being pulled apart by convection in the upper mantle below. The space they leave behind is where new ocean crust is born as it upwells because of decompression melting. The fresh crust is made almost entirely of basalt, and you'll recall that this is because basalt is the rock that forms when we melt the peridotitic mantle.

The speed at which the plates move apart from one another can be anything from 10 millimetres (0.4 inches) to 170 millimetres (just under 7 inches) per year. Although these values might not sound like much, particularly in relation to the size of the planet, if we calculate those distances over geological time, the results are spectacular. Bathymetric maps of the undersea world, which show the depth of the seafloor and its topography, clearly show a network of ridges as a scar around the planet where plates diverge. For the overall mid-ocean ridge system to accommodate the roundness of the planet it finds itself on, it is offset at various points by faults, allowing the ridge to make a side-step around the curve of the Earth. The ridge looks like the attempts Earth has made to heal itself after being ripped apart – which is essentially what is happening – but these 'scars' can really be thought of as a single 75,000-kilometre (45,000-mile) volcanic chain: the longest mountain range in the world. The thing about this line of volcanoes is that they hide, on average, 2,600 metres (8,500 feet) below the surface of the Earth's oceans. It is possible that there are over one million submarine volcanoes, tens of thousands of which rise over a kilometre above the ocean floor.

The spreading ridge is characterised by the highest mountains and the deepest canyons found on Earth, but the general appearance of the ridge in any one location is controlled by how fast that particular region is spreading. So, as you move to different parts of the ridge, it shows a different structure and style of eruption as some parts shift apart quicker or slower than others. Either way, even though these underwater volcanoes may account for around 75 per cent of the annual output of magma on Earth, they have rarely been observed to erupt because of their relatively inaccessible locations in deep oceans where the pressures are crushingly high, with the weight of the water giving pressures some 250 times greater than atmospheric pressure. Coupled with this, visibility is usually extremely poor. As a result, in the mid-Atlantic, scientists have mapped less than 10 per cent of the ridge in detail, suggesting that there is still a lot to learn about these important magma production centres.

Nevertheless, technological improvements in recent years have allowed a few eruptions to be observed, even in some very deep waters. Interestingly, the conditions found at the bottom of our oceans are not all that different from the environment on the surface of other planets and moons. This partly helps to account for the reason that we haven't been able to explore many of these alien surfaces in detail yet. After all, we've barely started to explore the more inaccessible and environmentally challenging parts of our own planet, so exploring similar ones in space is obviously even harder. Building scientific equipment that will work in these seemingly unearthly conditions is challenging and getting it there is far from easy. Most of the exploration of our deep oceans is carried out by robotic instruments, which have their own drawbacks. However, designing equipment to support humans in these inhospitable environments is extremely challenging and expensive so a lot of deep-space and deep-sea exploration requires that we make do with robots.

As lava erupts at a mid-ocean ridge, the way it flows is a result of the strange (perhaps only to us) submarine environment in which it finds itself. In fact, the lava doesn't even flow in this setting, simply because it is unable to. The lava at mid-ocean ridges can mainly only erupt in an effusive manner – meaning it is not explosive. Deep under the weight of the frigid ocean, lava tends to extrude out of the Earth's crust rather slowly, like squeezing toothpaste from a tube. This is in direct response to a number of factors. One of them is related to volatiles, or gas, within the magma. As we've already seen, the mantle material in these regions is less volatile-rich than the mantle at a subduction zone, for example, and it also doesn't tend to sit around at depth for very long before eruption, meaning it doesn't have as much time to build up gases. Without the volatile material, the magma stays at a lower pressure over time, making it less likely to explode. Furthermore, even if the magma contained a lot of volatiles, they would struggle to exsolve out of it because of the high-pressure setting the lava finds itself in at the bottom of the ocean with the weight of water above.

The lava could be described as oozing out of the mantle in response to the lithospheric lid above being gently pulled away. As a result, the outer surface of the extruded lava cools rapidly on contact with the very cold water it meets in the deep ocean because of rapid heat transfer by convecting water, and so forms a glassy smooth skin through a process known as quenching, which is simply caused by rapid cooling. This quenching turns molten silicate rock straight into silicate glass, producing a brittle rind on the lava. Structures known as pillow lavas are formed, so-called because the bulbous lobe of lava with its outer quenched surface continues to inflate from the pressure of new lava flowing into it, forming a rounded 'pillow' shape. Eventually the outside of the lobe cracks under the pressure and a new pillow forces its way out, flows a little distance and starts to form a fresh lobe of lava. The process of forming pillows is almost continuous, creating thick stacks of them on the ocean floor. Some researchers suggest that pillow lavas might be one of the most common types of lava over Earth's history, as spreading ridges are responsible for a huge portion of our planet's magma output.

## The meeting of fire and ice

Pillow lavas don't just form under oceans; they can form anywhere hot lava is erupted and meets something very cold that can remove the heat and promote rapid cooling. It shouldn't then be a surprise that they have also been found under lakes and even glaciers, where subglacial volcanoes erupt and are trapped by ice. While they are probably more commonly formed in water, they can even form in wet sediments.

Sure enough, when lava erupts under a glacier, it melts the ice, but if the eruption rate is relatively low, the water is still cold enough to quench the outer surface of the lava and form interesting pillow shapes. But when the rate of eruption is higher, the lava can melt its way through the ice completely, creating a more recognisable lava flow and with it, a flood of water. This is exactly how the jökulhlaups, or melt-water floods, discussed previously, are formed and become a dangerous volcanic hazard.

I'm describing the formation of pillows lavas on Earth in some detail because they are also important when we move into space. It has been noted that the slopes of Arsia Mons – one of Mars' extinct giant equatorial volcanoes – are marked by scars indicating that it had once been buried under piles of ice that moved and scoured its surface. Coupled with this, using data from the Mars Reconnaissance Orbiter, researchers discovered the existence of features on the surface in the same region that look just like pillow lavas. They also found other ridges and mounds that could have formed when lava was restricted beneath a glacier.

These findings are important because, as we've seen with the jökulhlaups, if we have ice and hot lava in close proximity then there is a high potential for melt water. While on Earth we mainly view these as a hazard, in space it might mean something completely different. With the production of liquid water comes the potential for the formation of a life-hosting environment. After all, scientists are quite certain that if we want to search for life in the Solar System, or even in exoplanetary systems, looking for water is a key starting point. Life and water are inextricably linked on Earth, but it is water in its liquid form that we think is required as opposed to its icy solid counterpart. A liquid solvent is capable of dissolving substances and enabling key chemical reactions, including in animal, plant and microbial cells, to occur. From studying just two specific lava formations on Mars' Arsia Mons, researchers calculated that there could have been two corresponding volcanically produced glacial lakes holding up to 40 million cubic kilometres of water, enough to fill twelve lakes the size of Windermere in the Lake District. Of course, just because there was potentially liquid water in this location doesn't mean life is certain to have occurred. Mars would also require the basic organic building blocks to have been present in the first place, and for the water to have persisted for a reasonable length of time for life to form and take hold. Nevertheless, because it is extremely challenging for scientists to search for microbes, dead or alive, without being on the Martian surface – either with humans or robots – knowing roughly where to aim a search in the future certainly helps, and this is one key location.

## Transatlantic communications and continental drift

Probably the most famous part of Earth's spreading ridge system is the Mid-Atalantic Ridge, separating the North American and Eurasian plates. I recall learning about this from a young age, but not because of its importance in the production of new oceanic crust. Instead, I was most fascinated and slightly amused by the fact that telegraph cables between Europe and the United States trail across the entire ocean, draped over these underwater volcanic mountains, with some parts of the ridge rising to 4,500 metres (14,600 feet) above the seafloor. In fact, the telegraph cables are so precariously placed that sometimes as new eruptions occur, they can get broken. It's not such a problem today as we have lots of very strong, and now fibre-optic, back-up lines protected by thick helical steel wire coatings. However, broken cables were more of a problem in the early days of transatlantic communications that used weak copper wires. Just laying the first cable across the Atlantic took three attempts, with success finally occurring in 1858 and the first message being sent and received that same year. Even so, it took over 17 hours to transmit the message because of terrible reception and a slow transmitting speed.

At the time of the first transatlantic telecommunications, the existence of the Mid-Atlantic Ridge was not completely known. In 1855, US Navy Lieutenant Matthew Maury had suggested that an underwater mountain chain might exist, but it wasn't proven until many years after. As part of the HMS *Challenger* expedition of 1872, Charles Wyville Thomson confirmed the presence of a ridge in the Atlantic while searching for a new location to place more transatlantic cables. The region was investigated further by sonar in 1925, a technique that uses sound propagation underwater to detect objects, and the ridge was found to extend all the way to the Indian Ocean, but it wasn't until the 1950s that it was mapped in detail, also by sonar.

But that's not where the story ends. The mid-ocean ridge was to play an important role in helping scientists to figure out the tectonic history of our planet. Soon after the sonar surveys had mapped out the morphology of the ridge in detail – and in turn illustrated that the ocean floor was not a flat, featureless region of the planet, as had previously been thought – it was discovered that the ocean bed was characterised by alternating magnetic stripes. Magnetometers initially used by aircraft to search for submarines during the Second World War were adapted to be pulled behind ships during the 1950s and started detecting what appeared to be strange magnetic anomalies at the bottom of the ocean. As more work was done, these supposed irregularities were found to form a pattern on either side of the ridge, when viewed over many kilometres, showing zebra-like stripes.

Sure enough, the seafloor pattern was something very significant. In 1962 two British scientists, Frederick Vine and Drummond Matthews, and Canadian geologist Lawrence Morley working independently, suggested that these alternating stripes were directly related to the polarity of the Earth's magnetic field when the basalt rocks were erupted from the mid-ocean ridge. Tiny crystals of magnetite contained within the erupting basalt lava align themselves with the Earth's magnetic field on eruption. They act like little magnets, which is similar in concept to the simple test you might have carried out at school with iron filings reacting and aligning to the field of a magnet. Once the lava cools, a record of the Earth's polarity is frozen into the rocks. Apparently, the magnetic field had changed multiple times while new rock was created and moved away from the spreading centre, leading to a mirror image pattern. One set of matching stripes either side of the diverging ridge has normal polarity – in the same direction as the magnetic field at the present day – then the next set has the opposite polarity, reflecting a different time in history when the Earth's magnetic field was reversed.

We should take a brief step back, though, and look at how all this was discovered. Earth's magnetic field has reversed many times over the course of our planet's history, and this fact was first discovered early in the twentieth century in the Massif Central mountains in France, but the stunning proof comes from the mid-ocean ridges. While the reversal of the magnetic field is the result of an internal change within the Earth's core – within its inner dynamo that generates the magnetic field – the way we would recognise it on the Earth's surface is that north on our compasses would point to the south pole instead. It happens, on average, once every 200,000 years, but the actual time between reversals has been highly variable. Remarkably, scientists think Earth's poles have not reversed for around 780,000 years. But let's not worry about that for now.

The presence of identical magnetic stripes on either side of the ridge, becoming progressively older the further away they are from the central ridge, supported the idea that the ridge was indeed diverging. This had valuable implications for geological research into our planet's dynamics and was one of the major clues that helped confirm the existence of continental drift and plate tectonics, which had been suggested in the early part of the twentieth century by Alfred Wegener, a German polar researcher (1880–1930). Without the necessary proof, Wegener's provocative idea from 1912 that the continents had once been joined in different configurations in the past – evidenced by the seemingly neat jigsaw-like fit of the coastlines of Africa and South America before they drifted apart – remained unpopular in many circles.

The new evidence about a spreading ridge snaking its way around the Earth led to scientists being able to develop Wegener's theory after his death. They went on to propose that the mid-ocean ridge system acted a little like a conveyor belt, shifting newly formed oceanic crust away from the ridge over time, and thus moving the continents further apart from each other. At the other end of the plate, the oceanic crust that had become cold and dense was subducted under continents as part of the tectonic cycle. Plate tectonics provided an elegant model for understanding Earth's need to allow heat to escape from processes occurring in the interior.

It might be that Earth is the only place in the Solar System to accommodate plate tectonics, but it isn't the only place to host magnetic stripes. NASA's Mars Global Surveyor, equipped with a planetary magnetometer, discovered banded patterns of magnetic anomalies on the surface of the red planet in the late 1990s. The question is whether these stripes represent seafloor spreading on Mars, which in turn would suggest Mars has plate tectonics. While Mars has no inner dynamo at the present day, it still has a magnetic field, albeit one that is weaker than Earth's. However, it may have had a stronger field in the past, and one that switched ends just like Earth's. Added to this is the fact that Mars hosts three volcanoes within its Tharsis region, of which Arsia Mons, mentioned above, forms an apparent chain just like many terrestrial volcanoes that are the result of plate tectonic movements.

But the evidence is still not strong enough to say that Mars has advanced plate tectonics as on our own planet. It might be that Mars attempted to start plate motions but that it failed to get fully under way, perhaps because it cooled too quickly, with its mantle too soon passing the point where it could convect and pull plates around. So scientists can't be sure how Mars' magnetic stripes were formed, and it is possible that there are other explanations that can account for their existence. Just because they look like Earth's oceanic stripes that represent the formation of new oceanic crust doesn't mean that they necessarily formed via the same process.

## Pulling apart

Spreading ridges aren't just found beneath Earth's oceans, but they can also occur where continents are being pulled apart, the East African Rift, which we have discussed already, being one example. It extends over thousands of kilometres as Africa is, at this very moment, attempting to split apart. It's just doing it at a very slow geological pace. Rifting began around 30 million years ago – it is a truly drawn-out process – and if it is eventually successful, a new ocean basin will be formed. Not only that but a new large island will be rifted off, composed of parts of Ethiopia, Somalia, Kenya, Tanzania and Mozambique, making the African continent smaller while forming a new subcontinent the size of India.

Despite the geologically slow progress of rifting (it might be more fun watching paint dry), it has impressive results. The supercontinent of Pangaea, which existed 200 million years ago, once housed the land masses of what is now Eurasia, the Americas, Africa, India, Australia and Antarctica combined. Continental rifting split Pangaea into the continents we see today, with the newly formed space between the rifting continents being filled by oceanic crust.

While we've seen that most of the mid-ocean ridge system sits hundreds to thousands of metres below the surface of the ocean, there is one key example where this isn't the case, and it is Iceland. While Iceland does indeed sit atop the centre of the spreading Mid-Atlantic Ridge, it just so happens that it isn't, in fact, formed entirely from the magmatism occurring at the ridge itself. The creation of Iceland is thanks, in part, to the presence of a mantle plume within the Earth. It is just a coincidence that that plume has found itself co-located with a spreading ridge. The reason the mantle melts to produce the huge volumes of lava that have piled up to create Iceland is because of two processes acting in tandem; decompression melting from the ridge and heat conduction from the mantle plume thermal anomaly. As we move away from Iceland, both north and south, the Mid-Atlantic Ridge returns to its normal state of a spreading ridge sitting deep below the surface of the ocean, because the influence of the mantle plume tapers out. We'll learn more about mantle plumes in the next section because they can tell us a great deal about the interior of our planet and some of the oldest portions of mantle it contains. Iceland provides me with the ideal link, because it's the one-off special case where the two different geological phenomena – mid-ocean ridges and mantle plumes – coincide.

## Visitors from the deep

Mantle plumes are an important source of volcanism on our planet, but they are probably an even more important source of volcanism on other planets, particularly the rocky ones, since they are the main way it happens. On Earth, mantle plumes produce a number of phenomenally beautiful and special island chains such as Hawaii and

Galapagos, to name just a couple. The thing about these islands is that their formation is almost completely unrelated to the surrounding plate-tectonic environment as they don't need a plate boundary in order to exist. I say 'almost' because, if it weren't for the plate motions, then the chain of islands that characterises these locations could not form. As we've seen in Chapter 1, chains don't form as a result of mantle plume volcanic activity on other planetary surfaces, because those planets don't have plate motions to pull the plate over the stationary mantle plume (Olympus Mons on Mars is an example). As such, the magmatic products of mantle plumes are often referred to as 'intraplate volcanism' as they are most commonly found within the middle of a plate, such as Hawaii in the Pacific. Because mantle plumes can exist regardless of whether plate tectonics is taking place, they are a key research focus for scientists who are studying volcanism on other planets, where plate tectonics is not known to be active and so their volcanoes are most likely to be related to plumes. By studying our own examples – which are slightly more accessible than those on other worlds – we can apply our findings elsewhere in space.

On Earth, mantle plumes may rise all the way from the core-mantle boundary, or they may originate in another location within the upper or lower mantle. The same is true for the other planets, but the problem in all cases is that we don't know for sure where the mantle plumes begin. Some scientists are even sceptical as to whether they exist at all. The issue is that we can't 'see' inside the Earth, or any planetary body, to prove whether these apparent chimneys of hot rock exist. Fortunately, there is some promising evidence that plumes do indeed exist, and that they could originate from the very deepest portions of the mantle, even if we can't view them directly.

Firstly, studies that have mathematically and thermo-chemically calculated the temperature of the magma that produced the mantle plume rocks at the surface show that mantle plume magmas are hotter, and therefore less dense, than the surrounding mantle – hence the 'hot' in the term hotspot. This helps to explain how they are capable of buoyantly rising to the surface from wherever they may start out. But in order for them to be so hot, they almost certainly didn't form in the upper mantle because temperatures here are known to be cooler than the deeper parts of the mantle, or core. These results usually show that the volcanic rocks produced by mantle plumes formed at temperatures a couple of hundred degrees higher than those calculated for the rocks produced at mid-ocean ridges. This is important because the volcanic rocks at mid-ocean ridges are said to have come from the 'ambient' upper mantle, i.e. the part that is representative of the normal average background. If rocks are found to form in environments hotter than this, then the explanation is that they are thought to come from the deep.

As we have already discussed, mantle plumes are just another way a planet chooses to cool itself by releasing its internal heat into space. A planet's internal heat is what drives any activity in its mantle, and it will help to produce the volcanoes on its surface as the molten mantle escapes. It's all part of a temperature-balancing act – a bit like a massive lava lamp. On Earth, the idea that mantle plumes may initiate and rise from the very deepest portions of the mantle is directly related to the fact that there is a strong temperature contrast between the outer core and the lower mantle. It is thought that the rock stored at the base of the mantle is heated by the core below until it reaches a point where it is warm, and buoyant, enough to rise. The thermal gradient for any planet will always range from hotter on the inside to cooler on the outside. So, once a 'warmer-than-average' portion of mantle starts to rise through a planet, there is no reason it shouldn't continue all the way to the surface, as long as it rises faster than it is cooled by the surrounding mantle.

These chimneys of rock rising through the mantle currently seem to be the best explanation for the hotspot volcanism observed on Earth, and in many places around the Solar System. These locations are all marked by anomalously hot and voluminous lava, compared to the ambient mantle. On Earth, the islands they produce show an age progression, as in Hawaii. While this is related to the direction of the plate movement upon which the mantle plume impinges, the volcanoes would erupt anyway regardless of the plate motion, and if there is no plate movement, end up piling lava upon lava. As we saw in Chapter 1 and here, this is exactly how scientists explain how Mars' Olympus Mons grew so enormous. The best explanation is that the feature feeding such volcanoes with fresh magma is stationary in the mantle.

However, there is increasing evidence that we can 'see' mantle plumes within the Earth using seismic waves, the energy released by earthquakes. We're going to cover this technique in more detail in the following chapter because it is an incredibly powerful way to find out about the interior of planetary bodies. However, put simply, it's similar to medical imaging. Detection and measurement of seismic waves within a planet allows scientists to build up a picture of the state of the material – whether liquid or solid – that the waves travelled through. As seismic waves are released by an earthquake, they travel through the planet and can be reflected, refracted, bent and bounced around inside, thanks to the planet's different layers. This technique has allowed scientists to piece together an accurate image of where the various layers within the Earth are located because the waves reacted to these layers of different states in different ways. The locations of the inner and outer core, the lower and upper mantle, and the lithosphere can all be known because they exist in different states – some hotter, some cooler, some liquid and some solid – so their boundaries are clearly visible. Mantle plumes, as we've seen, are hotter than the ambient mantle surrounding them. However, the issue is that the entire mantle is 'hot' and that we're only really talking about the plume being a few hundred degrees warmer than the material around it, perhaps 1,500°C (2,732°F) rather than 1,200 or 1,300°C. Trying to detect and define the edges of this apparent thermal anomaly is difficult as it's not just a black and white, solid versus liquid distinction. Everything is a sort of grey, being composed of 'squishy' material, and therefore what we are really looking at is more accurately described as solid mantle with similar, but distinct, relative temperatures.

Nevertheless, by combining lots of seismic data from all around the globe, scientists have successfully managed to image the plume sitting below Iceland, in a process known as tomography.* Tomography is a useful tool that lets us see the structure of anomalies within the mantle, and it has been used to depict a region of rock around a few hundred kilometres in width sitting beneath Iceland, which extends down to at least 400 kilometres (250 miles). This technique allows scientists to see regions of anomalous seismic velocity (how fast or slow the seismic waves travel), which might be

---

* A technique for imaging the subsurface of the Earth with seismic waves. The data can be used to construct 3D images of features within the mantle that have seismic velocity anomalies because of variations in their structure, heat or composition.

due to temperature or composition. The researchers who carried out the work suggest that while they can't yet image the plume deeper than the upper mantle, it could easily extend deeper, but it is not possible for them to resolve this anomaly to lower depths with current techniques. But thankfully, techniques will improve over time and we will get better at this as we deploy more seismometers and refine techniques. The great thing is that we can combine this geophysical technique, which tells us about structures within the mantle, with geochemistry, which tells us about compositional heterogeneity, to teach us more about what is hiding down there. Basically, this allows us to see whether the region where a mantle plume appears to have been detected seismically also has a different chemical composition.

When studying the rocks produced by mantle plumes, their chemistry has a lot to say. I spent years studying rocks from the Iceland plume. The thing about this plume is that it is apparently very long-lived, and we have been able to trace its origins back over 60 million years to when it first rose through the mantle to meet the lithospheric lid of the planet. The so-called proto-Iceland plume didn't find the easiest spot to meet the Earth's lithosphere as it initially rose underneath Greenland and Scotland, years before Iceland itself existed. As such, it had to work its way through very thick and old continental lithosphere. Once it managed this, however, in a period of continental rifting, it began erupting especially large volumes of lava. In fact, the start of the Iceland plume eruption produced a Large Igneous Province, which is now known as the North Atlantic Igneous Province, extending from Scotland in the east all the way west over to eastern Canada and places such as Baffin Island. This sounds like a large area, and indeed it is, but it was actually a little smaller at the time because the continents of North America and Eurasia hadn't begun to drift apart from one another yet. The mantle plume didn't care what kind of surface it was going to meet, it just kept going, regardless of the thick tectonic plates initially on top. At the point where it eventually caused the plates to break apart, marking the start of rifting, the production of new seafloor began in the northern mid-Atlantic.

The point of this story is that I studied some of the gases that were trapped within these special plume rocks. These were gases that had been transported by the magma from the Earth's interior to the surface. In particular, I focused on analysing helium gas. The question I wanted to answer with this research related to where exactly the helium had come from. Obviously, I knew it was from the interior of the planet, but I wanted to see if the helium could tell me anything about the history of the region of mantle from which it came, deep within the planet.

You'll remember that I've referred to the upper mantle as 'ambient' because it represents the average background composition of the portion of the planet that melts to produce new oceanic crust at mid-ocean ridges. Because of this, it is relatively well mixed compared with other parts of the mantle because it is actively and more efficiently convecting. The upper mantle is also classed as well degassed because it is constantly releasing volatiles through mid-ocean ridge magmatism. On the other hand, scientists think that the lower mantle, sitting below 670 kilometres (420 miles) depth and extending all the way to the outer core, is the opposite; it is less degassed, because it doesn't appear to play an active role in surface magmatism, and it is also less well mixed because of its higher density, which makes convection within it much more sluggish. Because of this, the lower mantle is thought to better represent the 'primordial' composition of the Earth's mantle: the composition that Earth started out with.

I studied helium gas to tell me more about the planet's degassing history because this element comes in two different varieties, or isotopes, that allow us to trace processes within the Earth and discover information about the origin of the magma we find at the surface. Our planet began with a finite amount of stored gas that it collected up from the cloud of gas and dust which formed our Solar System, and all of the objects within it, 4.5 billion years ago. Gradually Earth has lost some of its volatile materials over time because, as we've seen, they are released via volcanism. We can study this process using measurements of helium.

The only difference between the two isotopes of helium that we'll discuss here is their masses, which reflects the number of neutrons in their nucleus: helium-4 has one extra neutron in its nucleus compared with helium-3. Helium-3 is said to be 'primordial' because it has been contained within the Earth since its formation, becoming trapped within the planet. The abundance of helium-3 in the Earth, therefore, only decreases with time as it is lost via magmatism at the surface. Because the upper mantle plays an active role in all volcanism on the planet, it has a lower proportion of this primordial helium-3. The lower mantle, on the other hand, being less degassed, retains more of its original complement of primordial gases.

Now we move on to helium-4, which is quite different. It is produced as a radioactive decay product of elements such as uranium and thorium. This means that the more helium-4 we find in a region of the mantle, the older that portion of Earth is as it had more time to build up the helium-4 decay product over time. Now you can see how helium can act as a clever tool for tracing the age of different regions of mantle. If we can measure the ratio of helium-3 to helium-4 in a rock, then we can start to understand more about the history of the region of the planet from which the rock originated. One of the isotopes has been there from the start and only decreases over time, and the other builds up gradually. Old, un-degassed reservoirs, which we would refer to as 'primordial', such as the lower mantle, will have a high ratio of helium-3 to helium-4 because they have not lost much helium-3, whereas the parts of the planet that are continually involved in mantle convection and volcanism should have a low helium-3 to helium-4 ratio.

This is where things get exciting, because when I measured helium within rocks produced by mid-ocean ridges, I found they had low helium-3:4 ratios. This was bang on what theory predicted because they come from the continually convecting upper mantle that has had time to release its helium-3 over time during volcanism at mid-ocean ridges. I was not the first to do this (such data had long been in existence), but I repeated some of those measurements as a check. Crucially, though, when I measured the helium in some of the rocks erupted as part of the early Iceland plume, specifically rocks collected as part of the earliest phases of the volcanism in Baffin Island and West Greenland, they recorded the highest helium-3:4 ratios ever measured on Earth. This was a very exciting moment in my research career, knowing that I was studying rocks that contained some of the most primordial material within our planet, the earliest trapped gas that had been inside the Earth for over 4.5 billion years. What this meant was that the region of mantle that had risen to produce the volcanism at the surface had probably come from the lower mantle. A mantle plume that had risen from the lower mantle, transporting some of the oldest regions of the Earth's mantle to the surface, fits very well with these findings.

When scientists have looked at other geologically similar locations such as Hawaii and Galapagos, where it is also thought that a mantle plume has produced the lava flows, they also measure high ratios. These aren't quite as high as those I measured from the Iceland plume, but they are definitely higher than those found at mid-ocean ridges, proving that they too didn't originate in the upper mantle. As with Iceland, the best way to account for these ratios is that the rocks produced at hotspots have been erupted by mantle plumes that have come from a reservoir in the Earth that hasn't degassed its helium-3. The only reasonable place for that to have happened is in the lower mantle. This means that mantle plumes could be sampling pockets of the mantle that are as old as the Earth itself and have remained otherwise untouched in over 4.5 billion years.

The evidence is overwhelming that hotspot volcanoes are sourced from the very deep portions of our planet, and this makes them very valuable scientifically. They act like a window into the parts of our planet that we can't access. We'll be exploring this concept further in the following chapter because scientists are always trying to find ways to learn more about Earth's interior, simply because they are unable to do so physically. However, as we've seen many times already, researching the interior of Earth helps us to understand the planetary bodies around us. Venus, for example, is thought to host nine mantle plumes that almost certainly rose from its core–lower mantle boundary, like those on Earth. These plumes have produced the crustal plateaus and volcanic rises that are a key feature of Venus' surface. We won't be getting a sample of these rocks any time soon, since no missions are currently planned to land on Venus. Yet we can make assumptions about its interior because we can see the products of the plumes at the surface, as long as we have a radar to see through its thick atmosphere.

Mantle plumes also play an important role in the construction of Large Igneous Provinces. The production of so-called supereruptions, mentioned in Chapter 2, and the source of these huge outpourings of lava and associated magmatism are thought to be related to mantle plumes. We still don't know the reason why some mantle plumes have apparently been more potent than others. It would seem they are capable of producing large volumes of lava in short timeframes and can be active for very long periods of time, or they can be happy to fizzle away quietly with just a small ocean island to show for their work. Is it that there are plumes of different sizes within the mantle, with some large ones that can produce supereruptions and some small ones that just produce a minor island chain? That is certainly a possibility, but in addition some research suggests that while a single large plume may rise from the base of the mantle, it might split up into smaller blobs of plume as it approaches the surface. One large lower mantle plume may, therefore, be capable of birthing a whole host of baby plumes in the upper mantle that fuel seemingly unrelated magmatism over a large region of the globe at different times, reflecting when that individual baby plume eventually reached the surface.

There is even a suggestion that the production of mantle plumes within the planet's interior is a cyclical event, a result of the mantle's balancing act of returning to the surface the mass of material entering it at one of the ubiquitous and long-lived subduction zones in a 'what goes down must come back up' scenario. The subducted portions of lithosphere that enter the mantle at subduction zones don't themselves melt to form the volcanoes we see at the surface, or not directly anyway. These lithospheric slabs help to promote melting in the part of the mantle they descend into. Some of these slabs of rock are even thought to break off and sink all the way through the mantle, like a leaf sinking to the bottom of a swimming pool. If these slabs find themselves piling up at the base of the mantle, and are heated by the core, could they be the source of mantle plumes? We don't know, but there is every possibility that they are.

## Moving together

I think it is safe to say that Earth's subduction zone volcanoes produce, arguably, our most impressive volcanic features. Let's just take the example of those of the 'Ring of Fire'. While not a geological term, the 'Ring of Fire' refers to the volcanoes that surround the Pacific Ocean, marking the boundary of the Pacific plate with the others it contacts. It stretches for 40,000 kilometres (25,000 miles), from New Zealand up through Indonesia, the Philippines and Japan. It then curves across the Aleutian Islands and down the coasts of Alaska, Canada, the West Coast of the United States and all the way to the tip of South America. These locations are marked by steep-sided, often snow-capped volcanic mountains that tower over the surrounding landscape.

The processes that take place to form these volcanic mountains are impressive in themselves. Here we find that two lithospheric plates are converging and one of them is being drawn down towards the centre of the Earth in the process. In these locations, the dense and old portions of the oceanic Pacific plate slip below another section of lithosphere, which can be oceanic or continental, in the process of subduction that we have already touched upon.

As the plate descends into the mantle it promotes melting in the part of the mantle lying above it, creating magma that eventually rises to produce lines of volcanoes sitting parallel to, and near to, the plate margin. This is the reason they are also included in this chapter, because new land is formed at the surface. The volcanoes that are produced are referred to as arc volcanoes, because of the arc shape they form when viewed from above. They can be either continental or oceanic arcs depending on which type of plate they sit on, but they often form the classic conical-shaped composite volcanoes, which are usually the first to spring to mind when we think of volcanoes. The name 'composite' stems from the way they are built up from sequential deposits of lava, tephra, pumice and ash (being a composite of these materials). They are the most common type of volcano above the sea on Earth, with well over 400 on the Ring of Fire alone.

A volcano erupting on one side of the Pacific cannot trigger one in another location, nor can an earthquake. Eruptions are random events related to the complex set of different tectonic boundaries in a particular region. Sometimes it seems like there is a flurry of activity within a few weeks or months, but often at these times the Ring of Fire is no more active than normal, it's just that the media caught on to one event, and then another, and started linking them. The downside of this is that it can scare the public into thinking the Ring of Fire is more active than normal, or in more extreme cases, that the world is going to end and we're all doomed. It may sound ridiculous, but poor-quality reporting of these events really can cause unnecessary concern amongst the populations located nearby.

Despite this, the volcanic activity that surrounds the Pacific is fascinating, partly because there is so much of it above ground that we can see. As we saw in the previous chapter, these subduction-related volcanoes are the ones to blame for many of the more wicked phenomena associated with eruptions, namely huge ash clouds and pyroclastic flows, as well as violent earthquakes. As we've seen, it's all thanks to the high silica content, along with high volatile content that makes the lava more viscous and stickier, and therefore more explosive. But this high viscosity is the thing that allows them to grow into the impressively high, fiery mountains they become.

As we saw previously, the water carried down by the subducted plate plays an important role in the magmatism of these regions. Without the water, there is every possibility that melting of the mantle wouldn't take place at all. The addition of this important volatile, along with a few others such as carbon dioxide, acts like flux at a foundry, promoting melting at lower temperatures than would otherwise be possible. We learnt in Chapter 3 that mantle peridotite does not melt of its own accord. It needs an extra push such as some added volatiles (the salt to the ice), or an extra source of heat. When we add water at subduction zones, however, it unfortunately also acts to make these already silica-rich magmas even more explosive.

Despite all their wicked ways, subduction zones are an important feature on Earth. They are responsible for consuming our oceanic crust to keep plate tectonics moving, and for creating our continental crust. They also modulate our climate, transporting water and other volatiles into the Earth's mantle. Plate tectonics and subduction are key to forming a varied surface geology and allow Earth to host two main types of crust – oceanic and continental – unlike any other planetary surface we know. Continental crust allows for landmasses above sea level, upon which humans and many other lifeforms that require a dry world have had the luxury to evolve. Oceanic crust lines the base of Earth's important oceans, allowing a world of water-based lifeforms to thrive. Without water, life probably couldn't have begun, and without a sustained supply of liquid water at our surface, life wouldn't have continued.

Continental crust is simply a more advanced type of planetary surface. In fact, we could look at the Earth as a more evolved version of the other terrestrial planets. Because Earth has developed and maintained plate tectonics over millennia, it has continued to change, forming different types of crust, whereas the other planets are stuck in an earlier era of evolution with simpler – although still not fully explored and understood – surfaces. In Chapter 5 we'll explore how Earth's continental crust forms and why it's unique in the Solar System, while looking at how other planets developed their own crustal surfaces, all of which are also unique.

# CHAPTER FIVE

# Creating a Life-giving World

It is understandable that our volcanoes, and our planet's active interior, have played an important role in the habitability of Earth. Volcanoes have released the gases contained within the Earth and used them to make an atmosphere that has supported the existence of life. But it's not that simple, because we know other planetary bodies have also had active volcanoes; indeed some are still active at the present day, yet we haven't found life anywhere else. The question is: why does Earth have such a welcoming atmosphere that supports life when the other planetary objects either have no atmosphere at all, or one that is so thick it's toxic?

Our atmosphere makes up less than 0.0001 per cent of the mass of the Earth, yet it is incredibly important, particularly for any forms of life attempting to make our planet their home. The surface pressure our atmosphere affords us allows water to exist on our surface, and it keeps us warm by trapping in Earth's blanket of atmospheric gases: namely, nitrogen and oxygen, plus water vapour, carbon dioxide and methane amongst, others. Without our atmosphere we simply wouldn't be here because apart from providing air for us to breathe, it shields us from radiation damage and controls the temperature of our surface. Earth would be a very cold place at night without its atmospheric blanket and very hot during the day, both of which would have major ramifications for life. For comparison, on the Moon, temperatures get to over 100°C (212°F) where the surface is exposed to sunlight, and fall below -170°C (-274°F) during the night; and Mars, with its very thin atmosphere and quicker rotation, still has daytime temperatures of up to 20°C (68°F) but night temperatures get down to below -70°C (-94°F). Sure enough, maybe some of the Earth's current inhabitants could survive in these alien temperatures, but we humans probably couldn't. We are a finicky species that can really only survive en masse within a very narrow temperature range.

One of the most warming of the atmospheric gases is carbon dioxide, creating the greenhouse effect; it has also got a bad name for itself in recent years as one of the most powerful of these greenhouse gases. Earth's mantle is packed with carbon and it is the fourth most abundant element in the Universe. Without it, there would be no life here. When volcanoes erupt, they inevitably shift the storage of this element from the interior of the planet to the exterior. Carbon is one of the most important elements required for keeping our planet perfectly warm for life. But just a little too much carbon dioxide can create a planet that is too hot for life.

It is generally accepted today that carbon dioxide composes between 10 and 40 per cent of volcanic emissions, although this percentage varies from volcano to volcano. A single plume from just one volcanic eruption can release millions of tons of carbon dioxide in just a few hours, yet even the rare and very large eruptions of recent years such as Mount St. Helens in the 1980s and Mount Pinatubo in 1991 only released around 10 and 50 million tons of carbon dioxide respectively. This might sound like a lot, but the volcanic effect is incredibly small compared with the global budget, which is controlled by human emissions, themselves standing at 29 *billion* tons per year. It would take a few eruptions the size of Mount Pinatubo's 1991 blast to occur every day for volcanoes to even get close to matching humanity's level of carbon dioxide release. Even if you take all the volcanoes, including those that aren't erupting but are still 'degassing' on a daily basis, it adds up to less than 700 million tons of carbon dioxide released each year. Remember, we're comparing millions and billions. The reason our climate has warmed since the late 1700s is directly related to the influence of humans, so we really can't blame volcanoes for the long-term climate changes, however large their eruptions.

Earth's early atmosphere probably formed not long after the planet first coalesced into a planetary body. Soon after, as it was cooling down from a ball of magma to a planet with a solid surface, it released volatiles from its interior. The early atmosphere was most likely dominated by carbon dioxide with smaller amounts of water, carbon monoxide, methane and ammonia. Oxygen was unlikely to have been a major component of our atmosphere at first. Nevertheless, without this early oxygen-free atmosphere – however toxic it may sound for humans and much of life on Earth – the chances are that life wouldn't have taken hold here at all. As the Earth began to cool, the water vapour released by volcanoes didn't remain in the atmosphere for ever; it eventually condensed to form surface oceans. These provided a perfect breeding ground for Earth's early organisms. However, scientists think that the reason Earth was able to condense water out of its atmosphere to make oceans is that it cooled down at just the right speed, quickly enough to form a cool, solid and stable surface upon which the newly condensing oceans could grow. In turn, the formation of our oceans played an important role in removing the planet-warming carbon dioxide from our atmosphere. During the formation of carbonate rocks in oceans, carbon dioxide is efficiently locked up in the shells of organisms that have died and become incorporated into these sedimentary rocks. This is an effective, yet geologically paced, way of counteracting the greenhouse effect to create suitable global temperatures that allow for the rise of other organisms.

Of course, the greenhouse effect is an important natural process that acts to keep our planet at a habitable temperature. Nevertheless, if we alter the amount of carbon dioxide in the atmosphere by too much, the greenhouse effect can get out of hand and create inhospitable conditions. The analogy we could use to understand the effect is a car sitting out in the sun. Everyone knows that the car heats up very quickly, even on a sunny yet cool day. This is because the light, or rather the visible radiation, to be specific, enters the car through its transparent windows. There it works on heating up the insides of the car such as the seats and dashboard. These materials re-emit some of this heat as invisible infrared radiation and the problem comes about because glass reflects some of this radiation as it tries to leave via the windows, thereby trapping it in the car. The result is runaway heating – the greenhouse effect – and the car gets hotter and hotter. Earth's carbon dioxide in the upper atmosphere creates the same issue. The re-radiated infrared heat from the Earth's surface, being at a different wavelength to the visible light that shone down, gets absorbed by the carbon dioxide as it tries to escape to space, and this layer then acts to reflect radiation coming from below. Therefore, heat remains trapped within the carbon dioxide blanket and acts to increase the atmospheric temperature.

## The effect of water

You might be wondering about water vapour on Earth, since it forms such an important part of our Earth system, allowing life to exist here, and is the most abundant gas to be released by Earth's volcanoes, constituting up to 60 per cent of emissions globally (although with wide local variations). As a greenhouse gas, water vapour also plays a role in global climate. The problem is that scientists are still unsure just how much of an effect it has.

One of the problems is that water is a condensable gas, as opposed to carbon dioxide and other greenhouse gases. This means that the amount of water contained in the atmosphere is itself controlled by the temperature. As the planet gets warmer, evaporation increases and so the atmosphere holds more water. By contrast, carbon dioxide, for example, doesn't condense at our planet's atmospheric temperatures, so it always remains in the gas phase. With more water in the atmosphere, temperatures increase because the clouds that are formed absorb and trap thermal energy radiated by the Earth's surface, so the Earth gets hotter. It is a positive feedback loop. But at the same time, clouds can prevent the Sun's thermal energy getting to the planet in the first place. However, how much of a cooling effect this has is variable and hard to quantify.

You'll see that the story of water on our planet is complicated and I am only summarising a small part of it. What is important here is the effect volcanoes have. So, let's have a look at the volcanic emissions to see if they are significant in Earth's overall water budget.

Researchers have studied recent eruptions using satellite data to track the amount of water vapour released by eruptions and how long it remains in the stratosphere. They discovered that while local increases in the ratio of water vapour to dry air can be increased considerably following an eruption, the effects are short-lived. Therefore, the researchers concluded that the contribution of water vapour released by moderate-sized volcanic eruptions is not high enough to have any effect on the long-term trend of global warming.

## Earth's neighbours

Due to its proximity to us, Venus is often described as 'Earth's evil twin', being the planet most similar in size (in terms of mass and radius) and chemistry to Earth, but one that has a hellishly hot surface, with average temperatures of 450°C (840°F), and a surface pressure of 90 bars – in other words, about 90 times the surface pressure on Earth. The reason Venus is so hot is not, as you might think, because it is closer to the Sun. Venus' 'hot enough to melt lead' temperatures are certainly more related to its uber-greenhouse atmosphere, which means that water cannot exist on its surface. Without water, carbon dioxide cannot be removed from the atmosphere by oceans as it can on Earth, thus creating a runaway greenhouse effect.

Venus has 1,600 known major volcanoes and potentially 70,000 in total. This is more than any other planet in the Solar System, and although Venus' volcanoes have not been observed to be active at the present day, there is some evidence that they might be and that we just haven't seen them directly. The emissions from Venus' volcanoes are part of the reason the second planet from the Sun has a constant cover of dense and toxic clouds. Venus hosts a very thick and toxic atmosphere, consisting almost entirely of carbon dioxide, but with corrosive droplets of hydrochloric acid and very little water vapour, all produced by volcanic emissions. It's probably not a surprise we haven't found life on Venus' surface!

Scientists think the reason Venus didn't manage to match Earth in producing a long-lasting life-affirming atmosphere for organisms on its surface is because it didn't cool down quickly enough. Venus just wasn't able to balance its heat loss as efficiently as Earth. We'll learn more about this in Chapter 8. In the grand scheme of things, and knowing that the terrestrial planets formed over 4.5 billion years ago, it might not sound like a big deal if it took a few hundred million years for Venus to cool instead of a few tens of millions of years. We would expect Venus to have originally pumped out the same concoction of gases from its volcanoes as Earth because it formed in roughly the same portion of space, and so it is made of the same basic building blocks as our own blue planet. Therefore, we would also have expected Venus to have had water locked up within, which would have been released into its early atmosphere much like on Earth. While scientists don't yet know for sure whether Venus had liquid water on its surface in the past, they have suggested it might have been present based on the same kind of climate modelling they perform on Earth. If their Venusian climate models are correct, then it is possible that liquid water was in existence there for some two billion years, meaning that the seemingly most hellish of terrestrial planets could have been habitable during this time. However, we can't be sure because scientists don't think rocks from that period of Venus' history are still preserved on the planet. But they estimate that at the present day, Earth has 100,000 times more water than Venus, with our atmosphere being composed of 40 per cent water versus Venus' 0.002 per cent.

So, we have to question where Venus' water went and why? It turns out that over time, water might have evaporated from Venus' surface because it was unable to condense out to form oceans. It is thought to have split into its components – oxygen and hydrogen – with the hydrogen, which is very light, being stripped out of the atmosphere by the force of solar or electric winds. If Venus had had a protective magnetic field like Earth then it would have been shielded from such a process, but, alas, it seems it did not.

But once enough water was stripped away from Venus then its slower rate of cooling, and lack of oceans, meant that carbon dioxide continued to build up in its atmosphere, exacerbating the greenhouse effect and making the planet hotter. As a result, its crust has stayed too soft to permit the development of plate tectonics. Without a rock cycle, carbon dioxide will remain in the atmosphere as it is unable to be drawn down into forming the carbonate rocks that line oceans. Venus' 450°C (842°F) surface is rather life-limiting for the carbon-based organisms we know of. The features that allow for the right conditions for life to develop seem to be so fragile. It really is all about 'location, location, location'. Earth might just sit in the environmentally balanced sweet spot for the right conditions for liquid water, and therefore life, to develop – a true 'Goldilocks' planet.

What I want to highlight here is that volcanoes on Venus, have almost certainly played a pivotal role in its evolution. Earth was able to find a comfortable middle ground for greenhouse gases for much of its existence, whereas other planets experienced different styles and rates of volcanism. The effect of attributes such as a magnetic field and the preservation of liquid water are also very important in determining the likelihood of life on other planets.

Despite Venus being regarded as a low priority in the search for life, a new finding has thrown open the possibility that it might harbour organisms. It is highly unlikely such life is on the surface, but instead it might be living in the clouds of Venus' atmosphere where scientists seem to have detected the gas phosphine, although at the time of writing there is some debate in the literature as to the validity of these claims. Even so, it is worth discussing because of its importance if proven to be there.

Phosphine is a toxic and rancid gas that is broken down easily by oxidation in Earth's atmosphere. Here, our phosphine is continually replaced by organisms, including from the bowels of badgers and fish and even from penguin poo. But the amounts that have been detected at Venus are 1,000 times higher than on Earth. Being a rocky planet like Earth, Venus' atmosphere will also oxidise its phosphine, meaning that if the detection is real then something, or some process, must be replacing it. There is certainly no evidence for badgers or penguins on Venus and modelling suggests that volcanic activity releasing phosphorus, the feedstock for phosphine, would need to be 200 times greater than on Earth to account for the phosphine, which seems highly unlikely since we haven't yet seen it erupt (more detail on this in Chapter 10). After ruling out everything they could think of, including exotic chemical reactions, the scientists couldn't account for the production of phosphine through non-biological processes. Therefore, biological activity within Venus' clouds, where it is much cooler than on the surface, remains a possible mechanism to produce the phosphine that might have been detected.

Much like the methane in Mars' atmosphere being a potential biomarker for life (something that is detected that could have been produced by life), phosphine might now be an important gas to search for in other planetary atmospheres too. Future space missions will hopefully probe further into Venus' atmosphere to find out once and for all whether phosphine is truly present and how it got there, but there remains the possibility that it is indeed produced by life. For a world that was at first thought to house sweaty jungles with exotic lifeforms thanks to its constant shroud of thick cloud as seen from Earth, suddenly the possibility for life has returned to our intriguing next-door neighbour. If it's there, it would help to show us that life is not quite as fragile as we once thought and that it doesn't necessarily require all the conditions we currently think are necessary for it to thrive.

We still consider liquid water as a rather important ingredient for life to allow chemical reactions to occur, and, as we've seen, Earth is not the only planet to have, or have had, water on its surface. Scientists now think that our other neighbour, Mars, may also have had liquid water on its surface in the past. For Mars, the evidence is compelling, partly because the geological record is better preserved, and so scientists can be more certain that the red planet was once more of a blue planet. This has important implications for the existence of life in the past, and Mars has always topped the list of places where scientists want to go to search for life. It might even be that Mars has more water today than scientists have long assumed. Some of Mars' water was certainly lost to space, just like Venus', because Mars lost its protective magnetic field around 4.2 billion years ago. But much of its water is thought to have been 'locked up' in rocks and even turned into ice, which can now be found at Mars' poles. There is even a study suggesting the existence of an underground liquid reservoir of water. The implications for life in the more recent past, or even at the present day, are exciting if it turns out that liquid water is still viable on Mars.

To further demonstrate how fragile this relationship is between a planet and its atmosphere, we can look at the

Earth's Moon. Today the Moon has such a thin atmosphere that it can't even technically be classed as one; instead it is referred to as an exosphere. The molecules present are so few and far between that although they are gravitationally bound to the Moon, they do not behave like a gas. However, around three to four billion years ago, the Moon may have developed an atmosphere that lasted for up to 70 million years, and it's all thanks to its volcanic past. The volcanic activity was so prominent that it started to put gases into the lunar atmosphere faster than they could be removed. While most of the atmosphere was eventually lost to space, with lunar gravity being just too weak to hold on to it, some of these volatiles may now exist in polar ice caps on the lunar surface. Such ice may even come in handy for sustained human activity and exploration of the Moon in the future. The problem here was not the Moon's location, since it is in essentially the same spot as Earth, but its size. A planetary body that is too small cannot hold on to its atmosphere, therefore it will struggle to maintain liquid water, with important implications for the development and evolution of life.

## Methane on Earth and beyond

Of course, methane is another greenhouse gas that should be mentioned, because it is also released by volcanoes. However, it is one of the relatively minor gases, with Earth's volcanic activity accounting for less than 10 per cent of all geological sources, dwarfed by biological sources. Atmospheric methane on Earth can come from biological sources such as livestock, and it is also released from natural wetlands. Nevertheless, by studying the effects of methane on Earth, whether volcanic in origin or not, scientists have been able to apply their findings to the climates of other planets which, in some cases, has led them to suggest that we can't rule out life in these places.

Let's take Titan, a moon of Saturn, as an example. Titan's dense atmosphere is mostly composed of nitrogen (over 98 per cent), but it has appreciable amounts of methane too, just under 1.5 per cent. The problem scientists find is that this is *too much* methane. The methane in Titan's atmosphere, being just a simple hydrocarbon ($CH_4$), should all have been broken down to ethane by the ultraviolet in sunlight over timescales of around 10,000 years. The fact that there is so much methane in Titan's atmosphere, or even any at all, suggests it must be finding a way to replenish it. Step in: Titan's volcanoes! Or to be more precise, ice volcanoes. Scientists predict that these could represent a reasonable mechanism for releasing methane otherwise trapped within Titan's interior out into its atmosphere.

Sure enough, NASA's Cassini mission – which studied Saturn and its system, orbiting the planet for 13 years – imaged a region of Titan known as Sotra Facula, which appeared to be composed of three conical-shaped mountains that resemble volcanic edifices. These mountains appear to include features around them that look like the deposits of lava that has flowed from their summits. They've led scientists to suggest that they could be ice volcanoes (because they know the surface of Titan is too cold for molten rock to flow) and, therefore, that they might represent a source of Titan's replenished atmospheric methane. We'll learn much more about this in Chapter 11. Nevertheless, there are potential alternative explanations. One of them is that the methane comes from life on Titan, just like much of it does on Earth. However, since it is not yet proven whether life exists in Titan's hydrocarbon lakes, this is the less likely (but more exciting) possibility.

Looking at Mars, we have also found methane in its atmosphere, but the problem is that none of the instruments that have so far detected it have any way of telling where it came from. Mars is fairly similar to Earth, much like Venus. But knowing what we know about the red planet and its history – that it once had liquid water flowing across its surface – scientists think there is an even higher chance compared to Venus that it had life in the past. Interestingly, methane has been detected many times in recent years, with NASA's Curiosity rover having detected it frequently over the course of the mission, and it also appears that Mars' methane levels rise and fall with the seasons. The abundance of methane on Mars is very small, but the fact that it is there at all needs to be explained. Unlike Earth, Mars' volcanoes haven't been active recently enough to account for the methane and even if they had been, we would have expected them to pump out a lot more sulphur dioxide and other volcanic gases at the same time, which are not detected in the right relative abundances. Therefore, volcanoes seem an unlikely option for Mars' methane.

That leaves potentially only two mechanisms to account for the detection of the smelly gas on Mars: hydrogeochemical (water-rock) or microbial processes. Either one of these would be fascinating because it would signal that Mars is an active and potentially life-giving planet at the present day. For the former, a hydrogeochemical process, we would need a subsurface reservoir of water at a temperature where reactions could take place between the water and overlying rock. This temperature would possibly be as low as 30–90°C (86–194°F). This has real potential because Mars is known still to be a warm planet internally. It's probably not warm enough to fuel volcanoes, but it still contains heat that needs to escape into space. This heat could warm up a small subsurface reservoir of water, especially if it was insulated by a rock layer above. Within this reservoir, where water meets rock, reactions could then occur involving hydrogen, carbon and metals – all of which are readily available within the rock – that produce methane as an eventual by-product. The latter option, a microbial process, involves methane production by microorganisms and the location for this could be the same, since we think water is such an important ingredient for life.

Both hydrogeological and biological processes produce methane on Earth and so they are something we understand in detail. All we need to find is a location with the perfect conditions for either to occur on Mars at the present day. Indeed, the European Space Agency (ESA)'s Mars Express hinted at the detection of a higher concentration of methane over an area purported to contain subsurface water ice, but this doesn't rule out either process discussed above, since a geological or biological origin could still be possible. While we are yet to see direct evidence for these potential methane-producing processes, scientists are monitoring closely and tracking Mars' methane to see how it varies over time. We are sure to figure out the secret to Mars' methane soon and, in turn, if the red planet hosts life.

## Life on a spreading ridge

We've seen that volcanoes on Earth have released Earth's mantle gases over the course of millennia, which has helped to create a hospitable atmosphere for life. Yet we've also seen that Earth might be the only place in the Solar System where this has happened successfully, despite not being the only active world. If large, intelligent lifeforms existed on the surface of another planet or moon then we would probably have seen them by now. These other worlds either have atmospheres too thin or too thick for life to be possible on their surfaces, so far as we currently understand it.

There might be one exception to that rule though and that is Titan, one of Saturn's moons, which has an atmosphere just a little denser than Earth's, providing a surface pressure just under one and a half times that of our own. This certainly makes the surface of Titan sound somewhat hospitable, yet we still haven't found life there. This might be related to the fact that Titan's atmosphere is mostly composed of nitrogen and methane, and there is no water on its surface. But more on this in Chapter 11, because Titan is a very interesting moon and, despite its drawbacks, scientists are still hopeful that this little active world might have a bigger story to tell.

Nevertheless, this doesn't mean our search for life elsewhere in the Solar System is over, because certainly not all of Earth's organisms are found on the surface, breathing in our life-giving air. We have a wonderful array of organisms living in our oceans and it might be that this is where life first took hold on our now very fertile planet. While scientists think life on Earth started around four billion years ago, working out *exactly* when, where and why are big scientific questions that remain unanswered for now. Nevertheless, the reason scientists think the deep ocean, in particular, is a reasonable candidate for the location in which life arose is because it often marks the meeting of water and volcanic activity, most commonly in the form of spreading ridges, where two lithospheric plates move apart from one another. Here there is a link between Earth's internal heat – an energy source – and a large body of salty water. These two things in combination provide an environment that can support biological organisms. We could, therefore, expect to find the same conditions at any place where an active volcanic centre meets water, such as at the submerged parts of Earth's hotspot volcanoes. Fortunately, Earth is not the only place in the Solar System where volcanic activity exists in close proximity to a body of water. Even if we haven't yet found evidence for spreading ridges on objects in space – there being no confirmation of plate tectonics on any other planet or moon, which is required to create a spreading ridge – we know there are active worlds out there, where warm rock sits below liquid water.

One of the most important features of a spreading ridge, from the point of view of life, is its hydrothermal vents. These are fissures on the seafloor where seawater is geothermally heated and discharged to create what look like chimney stacks up to 13 metres (40 feet) high. They are a common feature on every mid–ocean ridge studied on Earth.

A hydrothermal vent forms when ocean water works its way down into the oceanic crust via cracks and holes, where it becomes heated by the mantle that sits just below. Where this happens at a fast-spreading mid-ocean ridge, such as the East Pacific Rise, the crust is usually thin and the mantle is, therefore, close to the surface. Consequently, the ocean water percolating down can be heated to extremely high temperatures, up to 540°C (1,004°F), relatively easily. Yet the water doesn't boil because it is held under such high pressure due to the weight of the ocean. Instead, the hot water collects up chemical species from the basaltic rock by leaching them out, and brings them up to the ocean floor as part of the cycle of hot waters through the system. When these waters reach the cold ocean water their dissolved minerals precipitate out rapidly on contact with the cool seawater to produce deposits of dark-coloured sulphides. This also creates what looks like a plume of smoke to form so-called 'black smokers'.

As we know, there are entire communities of organisms living under the sea. Many of these we already know well, such as the stunning fish, whales and crustaceans that inhabit the shallower regions of the oceans. But the organisms that inhabit the regions around hydrothermal vents and black smokers are less *The Little Mermaid* and more *Tremors* (the 1990s franchise of movies centred on strange subterranean worm-like creatures). The creatures that exist in the cold and crushing depths of our oceans remind me of the famous line from *Star Trek*, which is commonly misquoted, 'It's life, Jim, but not as we know it'.

Such spreading ridge biological communities were first discovered on the Galapagos Ridge in 1977 by *Alvin*, the deep-sea submersible. Compared to the majority of the deep sea, the regions close to these hydrothermal vents were found to be incredibly biologically productive. But the problem scientists instantly found was that there is no sunlight at these depths, so it wasn't obvious how these organisms survived. What we now know is that black smoker organisms are often known as extremophiles: they positively love extreme (that's extreme to us humans) environments, such as radically un-human temperatures (when they are called thermophiles), acidities (acidophiles), alkalinities (alkaliphiles) or chemical concentrations (chemophiles). While waters with temperatures of 350°C (660°F) might not sound very habitable to us, for many organisms they are bearable for short periods, even if the upper limit that any carbon-based life can survive permanently is said to be 123°C (253°F) – the temperature limit for maintaining stable carbon bonds.

The biochemical reactions that fuel the organisms on hydrothermal vents are collectively known as chemosynthesis, which involves the biological conversion of carbon-containing molecules and nutrients into organic matter.

The reactions use inorganic compounds such as hydrogen, hydrogen sulphide or methane as energy and combine them with an oxygen source, which is in the seawater. The major similarity that these deep ocean biological communities have to those that make their home above the sea is they require water. However, the major difference is that they don't require sunlight to make organic matter.

While mid-ocean ridges supply around 25 per cent of Earth's heat flux, by efficiently transferring its internal heat supply out into the oceans via magmatism, they also deliver chemicals from the rock to fuel these rare and unusual ecosystems. Here, life doesn't require sunlight because it is specially adapted to cope without it. Instead, microorganisms rely on the flux of Earth's geothermal heat and the dissolution of rocks to get access to the chemical species they need to catalyse reactions and create energy. However, while the species found at these depths don't require *sunlight*, it doesn't necessarily mean they don't require *light*. Some very special green sulphur bacteria capable of photosynthesis have been found on a deep-sea hydrothermal vent. These are anaerobic bacteria that require only exceptionally low light levels for growth by oxidising sulphur compounds to reduce carbon dioxide to organic carbon. Without sunlight, the bacteria instead use the light given out by geothermal radiation from the Earth at hydrothermal vents, which includes wavelengths absorbed by photosynthetic pigments of this organism. It is truly fascinating!

The microorganisms found living on the mid-ocean ridges make up a weird but important part of Earth's biosphere and form the base of the local food chain. They even support a diverse range of other organisms including giant tube worms, clams, limpets, shrimp and possibly even sharks. We mentioned earlier the Kavachi volcano in the Solomon Islands because of its repeated appearing and disappearing act as it erupts and then is once again eroded by the sea. While this is not a deep-sea hydrothermal vent, the undersea environment around the volcano has many similarities, in that it is not obviously hospitable to life; it's got a high level of acidity, and it's very turbid and warm. The discovery of species of sharks and stingray living within its underwater crater was therefore a surprise to scientists. Sharks have been spotted darting in and out of the clouds of active volcanic plumes, giving this volcano the nickname 'the Sharkcano'. It might be that the conditions here aren't great for the sharks, but the fact is that they are finding a way to live in this place, even if only for short periods.

Back to the hydrothermal vents: it's not just bacteria that survive down there. One of the larger organisms is the Yeti crab, so-called because of its very hairy legs. This blind species (after all, there's no need for eyes at these depths when there's no light) measures around 15 centimetres (6 inches) long and was first observed by marine biologists in March 2005 on the *Alvin* submersible along the Pacific–Antarctic ridge, south of Easter Island. The Yeti crab was living at depths of about 2,200 metres (7,200 feet); as yet, the scientists are still not sure exactly how it survives down there. While it was noted that the crab appeared to be eating mussels, which would certainly provide it with nutrition, the scientists also think that its hairy legs are home to large colonies of filamentous bacteria and that it might be 'farming' them for food. For now, it remains a mystery, but the Yeti crab shows us that life will thrive even in the most seemingly inhospitable of places.

The existence of populations of strange organisms on our own planet – those that don't require all the things we would normally consider to be necessary for life – opens the possibility for life elsewhere, including exoplanetary systems quite unlike our own. With this information in mind, planets far from the Sun, or even those with a subsurface ocean in which sunlight is blocked by a crust of ice, could have the potential to house some basic organisms. Such alien organisms could even exist in places that we might not initially think are biologically conducive to life, since we now know extremophiles can survive in environments that seem too extreme to us but for them are quite comfortable. Of course, anything within the Universe still has to adhere to the usual laws of physics, so perhaps we don't need to expect alien life to look that different from the organisms existing in the deep oceans on Earth, if they experience shared conditions in extremes of pressure and temperature.

You might be wondering where within the Solar System we could expect to find these environments. In 2005 NASA's Cassini spacecraft sent back images and data from Enceladus, showing that this little moon of Saturn was home to a subsurface liquid ocean. It was a tremendously exciting moment because, as we discussed earlier, scientists thought it should be frozen solid. However, the fact that Enceladus, while icy at the surface, was also shown to host salty, liquid water meant that, with its apparently warm rocky core, there was the potential for it to support life.

In 2015, Cassini flew through Enceladus' plumes of water, ice and dust that shoot out at 400 metres (1,300 feet) per second, emanating from its subsurface liquid ocean. Within these plumes Cassini's instruments detected not only sodium, proving the ocean is salty, but also nano-silica grains, molecular hydrogen and small organic molecules. Right, so it's not life. However, the nano-silica grains only form in water that is near boiling point, indicating that Enceladus' liquid ocean is in contact with a hot, rocky interior rather than another layer of ice.

The nano-silica is thought to form when water enters Enceladus' seafloor, becomes heated by the rocky core below and picks up silica from the rock, which then gets precipitated out when the heated fluids rise back up and meet the relatively cold ocean. This is just like the hydrothermal vents on Earth's mid-ocean ridge system, where nano-silica grains produce so-called white smokers, existing in the same hydrothermal regions as the black smokers. The molecular hydrogen was further proof that hydrothermal processes were taking place on Enceladus, providing a potential energy source for life. The organic molecules can be formed abiotically – not requiring life – but on Earth they are commonly formed biotically – by life. Scientists just don't know yet how those on Enceladus came into being, but they certainly can't yet rule out the existence of life.

Unfortunately, this is as far as Cassini got. When the spacecraft launched in 1997, scientists didn't even know Enceladus had liquid water or plume activity, so it didn't take along instruments to detect anything like organic molecules. The scientists hadn't even planned to fly through the plumes, because they didn't know they were there. The instruments that made these discoveries were designed to study Titan's atmosphere and Saturn's rings. The Cassini team learnt the age-old lesson of space exploration: expect the unexpected!

The result of these discoveries is that Enceladus might now be the Solar System location where we have the highest chance of finding life. Europa (one of Jupiter's moons) was always in the running, and still is, because it has a subterranean liquid ocean that is probably also salty, with a rocky core beneath that. NASA's Clipper mission will be on its way to Europa in the mid-2020s to investigate whether it harbours conditions favourable to life. However, Enceladus offers a potentially more inviting radiation environment compared to Europa, which is flooded by Jupiter's radiation. There are other mission proposals in the pipeline, so we will just have to wait patiently to see if they come to fruition for us to delve into these possibly alien-supporting worlds.

## Volcanic fertility

Having looked at the life-giving environments that volcanoes can create underwater on Earth, it would be good to focus on the uses they have above the water too, in addition to their support of our atmosphere. This section necessarily focuses on Earth, since we haven't found life elsewhere, so it won't apply to other planets unless we do discover life there one day.

We've touched on the fact that despite the fairly impenetrable nature of hard lava flows, organisms can inhabit them relatively quickly, as we saw earlier with the Hawaiian trees that rapidly return to inhabit fresh lava flows, and provide the breeding ground for more organisms to follow. Nevertheless, lava flows are still a difficult environment for many organisms. While humans can use the rocks from lava flows to build sturdy homes, they are a struggle to excavate because they are so hard, particularly in developing nations where industrial equipment is harder to come by. In addition, farming on land made up of young lava flows is all but impossible. However, when we focused on the more explosive types of volcanoes, we found that they are much more likely to produce large volumes of ash, and while these volcanic products can be life-threatening at the moment of eruption, they do have their uses once settled and cooled (if you managed not to get caught up in their ferocity).

You might wonder why people choose to live on the slopes of volcanoes. I know I have questioned it before when I see that a volcanic eruption somewhere on our planet is pending, or actively under way, and people are having to flee from their homes. Why did they choose to live there in the first place? While there are many socio-economic reasons for people choosing, or having, to live in these locations – perhaps because the slopes of the volcano provide affordable land to build on – one of the big pulls is the fertility of the soil and the lushness of the land surrounding these fiery, unpredictable mountains.

When ash rains down following a volcanic eruption, it can be a light dusting at one time and at others swamp entire towns. Obviously, the thick layers of ash are rather life-prohibiting, at least initially, as we saw with Pompeii. Big ash flows and fallout create more of a barren, Moon-like landscape, engulfing everything. However, whether a small or large amount of ash falls on to the land, it has some amazing properties. The thing about ash is that it is fine-grained, made up of small particles of exploded and cooled magma. As such, its large surface area compared to volume means that it is broken down quickly in the natural environment from the action of water, wind and even organisms. During this weathering process, the chemicals contained within the ash are released into the soil, acting as nutrient sources that make the land more fertile. Just like you might dig some organic matter into soil to help your plants grow, volcanic ash can add key chemicals such as phosphates, nitrates, potassium and calcium, among others, which aid plant growth. Ash also has the potential to hold a lot of water and the minerals it contains can bond easily with organic matter, holding it in the surface layers of the soil. As a result, the slopes of volcanoes that receive a dusting of ash, even infrequently, can be particularly productive. Hence, people often choose to live on land that is close to an ash-producing volcano for farming purposes, perhaps turning a blind eye to the potentially deadly aspect of it.

Researchers have studied the slopes around Krakatoa in Indonesia, along with those of Mount St. Helens in the United States, as case studies for how volcanic slopes regenerate following the upheaval experienced during and after an eruption. Krakatoa provides a longer timeframe – over 100 years – in which to study such recovery following a large eruption, versus the 30 years since Mount St. Helens erupted. While it's noted that these volcanoes inhabit very different climate zones, the results indicate that even Mount St. Helens' cooler subalpine slopes could be restored to their former glory within 100 years (minus a huge chunk of mountain that got blown apart during the eruption). Mount St. Helens could soon see the return of its majestic Douglas fir and western hemlock, just as Krakatoa's slopes are now covered with rich tropical rainforests, vegetation and wildlife. Such regeneration is all part of the cyclical biological renewal of volcanoes thanks to the natural fertiliser provided by the eruption. Life wants to cling on, so despite the eruption initially flattening the region to produce a seemingly barren expanse of land, we have to remember that life will find a way to return and thrive. In fact, on Earth it seems quite hard to stop it. We have to accept that even if humankind were to die out, other forms of life on Earth would almost certainly continue in our absence, specially adapted to the new environment. The question is: can we expect the same to have happened elsewhere in the Solar System? For now, we can only hope.

# CHAPTER SIX

# Peering In

We know that the terrestrial planets – Mercury, Venus, Earth and Mars – are all volcanic worlds, even if not all of them are active at the present day. But we also know that's not where volcanic activity in the Solar System ends. From the moons of Jupiter and Saturn all the way to Pluto in the Kuiper Belt, scientists have found evidence of volcanic activity. Much of it doesn't necessarily look like what we know so well on Earth, but it does an important thing: it indicates that the world we are viewing is active, or was in the past.

Volcanoes are particularly valuable scientifically because they dredge up the deep planetary interior and helpfully deliver it to the surface. In this way, they provide samples of the interior of these worlds that can't otherwise be accessed directly. But there is a problem. While volcanoes offer up such trinkets, these only sample a small portion of the total volume of that world. Yet, despite their small size, they do allow scientists to estimate the composition and structure of the interior of a planetary body, even if that is a notoriously difficult task. They are extremely valuable because so much of any planet is otherwise physically inaccessible. Nevertheless, sampling the very deepest portions of a planetary body, the core, is all but impossible (unless you believe the Hollywood movies, that is, for example the 2003 sci-fi movie, *The Core*). There is one exception, though, and that is iron meteorites. While these space rock samples are not from Earth, they are interpreted to be parts of the core of planetesimals* that were destroyed early in the history of our Solar System. They are just shards of a world long gone, representing a piece of the core of a shattered world that we can use to tell us about our own intact one. We will discuss them further later in this chapter.

As you can see from this discussion, the rocks volcanoes bring to the surface are really just a whiff of what's lurking down there. Scientists also have to be careful that the rocks are not changed on their long, and often complex, path to the surface, which is not exactly a quick elevator ride. Rocks brought up from the deep can be altered on their way, having been ripped from their warm home in the mantle and dragged up through the crust to the cooler surface. When scientists study them, they must do a lot of work to pull apart the processes that the individual trinkets of rock experienced on their journey to eruption. So, while volcanoes are useful for sampling planetary innards, they are just a small piece of the whole story. Pained as I am to admit it, to investigate in detail what's down there requires a much wider approach than simply studying the products of volcanoes.

You might wonder why we don't just drill into the centre of a planet; the reason is because it's impossible. In fact, the deepest we've ever drilled down is 12,262 metres (40,230 feet) in a location known as the Kola Superdeep Borehole in Russia. While the 12-kilometre (7.5-mile) distance is not to be scoffed at, this borehole penetrated just the upper

---

* Planetesimals are solid objects formed from dust, rock and other materials within the debris disc around a young star. A planetesimal can be anywhere in size from a few metres to hundreds of kilometres across. These objects in the Solar System are thought to have coalesced to form the planets. See also Chapter 7.

30 per cent of the crust in that region, not even touching the mantle below. It's barely a smidgen compared to the distance to the Earth's centre, which is over 6,000 kilometres (3,700 miles) from the surface, making it a completely inaccessible part of our planet.

So, how do we know what's down there? To be quite honest, we *don't* know for sure. Of course, we can collect rock samples on the surfaces of planets and from these it is possible to estimate the composition of the upper portions of that planet. For the Earth, this can take us down to around 200 kilometres (125 miles) in depth. But it won't take us the further 3,000 kilometres (1,850 miles) to the base of the mantle, and certainly not to the core below that. A further issue is that the planets and moons are internally layered, meaning they're not the same composition all the way from the top to the bottom. Collecting samples from the surface sometimes tells us almost nothing about the deeper portions of a planetary body.

This means that most of the knowledge we have about the materials that make up the very deepest portions of our planets and moons comes from indirect evidence. We might call this circumstantial evidence, if we were in a court of law. While direct evidence provides almost irrefutable proof of a crime, circumstantial evidence requires further inferences to be made for it to be useful. This is not dissimilar to the way scientists infer what is inside a planet. They have a great deal of evidence, some of it direct, but much more of it indirect. They use this body of information to build up a scientifically reasonable interpretation of what they conclude from this evidence to be down there.

For example, if the charge was that someone had hit a person with their car, then a piece of direct evidence could be an eyewitness account. This can be likened to our rock samples from the Earth's mantle. But one eyewitness account is often not enough to convict someone. In lieu of, or in addition to, witnesses, the court might rely on indirect evidence. This could be a dent in the car that was assumed to have been produced during the collision. Even if there were no eyewitness, further evidence – perhaps also that the defendant was driving under the influence of alcohol – would make a stronger case for the jury to decide whether the defendant was guilty of the crime. In the case of our investigations of the Earth's mantle, we might sample gases in tiny little bubbles in magmas at the mid-ocean ridges for our witness statement. A multidisciplinary approach is therefore required when scientists want to determine the depth and composition of the many layers of a planet.

Here we're going to look at these lines of evidence to prove to you, the jury, what is inside our planet, and how this applies to the other planets around us. Scientists and their disciplines will serve you as witnesses and barristers, providing you with evidence, and discussing the circumstances.

## Chunks of Earth

While most volcanic rocks at the surface of our planet are themselves products of the mantle, the magmas that produce them undergo changes on their ascent to the surface such that they no longer represent the mantle they left behind. Sometimes magmas contain an actual piece of the mantle: a rock that has been ripped from depth and transported to the surface intact. These little trinkets of crystallised mantle are called xenoliths, from the Greek for 'foreign' and 'rock', signifying that they are an extraneous component within the magma. This, however, makes them very valuable to scientists, who can use them to analyse and understand different parts of the Earth's interior to which they wouldn't otherwise have access. The problem is that these xenoliths formed under different pressure and temperature conditions at depth from the conditions they travel through to reach the surface. They sometimes also undergo a violent journey to the surface that doesn't leave them totally unaffected. Nevertheless, researchers are often able to back-track the processes these rocks from the deep underwent on their rise.

Despite their journey from possibly great depths within a planetary body, most xenoliths studied on Earth only sample the top few layers of our planet, allowing scientists to peel away just the crust and a bit of the mantle. Still, that is much further than we can otherwise access. The Earth's mantle makes up about 80 per cent of our planet, extending down to 2,900 kilometres (1,800 miles) in depth; it's a massive volume that is understandably difficult to sample.

Yet there are some very special rocks that can help us out in this quest. This is where the story of diamonds comes in. Despite the high price of diamonds, many people probably don't appreciate that the shiny rock they wear on their finger holds key information about the inside of our planet. Diamond crystals represent a little piece of the Earth's interior, having originated very deep down. They were formed under immense pressures and temperatures over one billion years ago, possibly even as far back as 3.5 billion years ago, at the base of the lithosphere, some 150 to 200 kilometres (90 to 125 miles) down. There is increasing evidence that some diamonds formed even deeper in the Earth's mantle, possibly at the transition zone between the upper and lower mantle, or in the lower mantle itself, which is over 660 kilometres (410 miles) deep. These unlikely messengers from the deep are actually hardy pieces of crystalline carbon. The crystals produced at these depths are so strong because each carbon atom they contain is covalently bonded to four others, making the hardest naturally occurring substance known on Earth. As we are told by the clever marketing of these stones: 'Diamonds are forever', which is true in a sense because they are such a hard and stable substance.

In the past, diamonds were brought to the surface by a special type of volcano called a diatreme or pipe, but these don't occur at the present day. We can think of a diatreme as an explosion below the surface that pushes magma, known as kimberlite, up towards the crust, as opposed to a classic volcano on the surface. Diamonds are not associated with any other type of volcanoes, only diatremes; this is just the way they hitched a ride to the surface, and the diatreme didn't itself produce them. Geologists can refer to a diamond as a mantle xenocryst, as opposed to a xenolith, simply because they are composed of a single mineral ('cryst') instead of a collection of materials that make up a rock ('lith'). But despite this, they can be just as useful as a more complex rock made up of lots of minerals.

Today, the remnants of such diamond-transporting eruptions can be found in places such as Australia, Canada and southern Africa, where there exist extensive diamond-mining operations to locate and remove the precious stones. These locations are all characterised by having especially thick and old lithosphere, regions called 'cratonic' (from the Greek for 'strength'). These are portions of strong lithosphere that have survived for a very long time, protected in some way from the processes of destructive plate tectonic cycles, usually because they are in the centre of a plate. They are geologically distinct and distinguished from other regions by being stable, with little or no seismic or volcanic activity, because they are so far from plate boundaries. Importantly, they have thick crust and an even thicker lithospheric 'root' extending several hundred kilometres deep into the mantle, like an iceberg in water. The root is an important factor in determining why diamonds are associated with these regions.

Without the special volcanoes that produced the diamond-bearing pipes that pushed up through the thick lithosphere, these precious crystals would have stayed trapped in the depths of the mantle indefinitely. Once diamonds are at the surface, they allow scientists to probe not only into the inside of the Earth, but also back in time to their formation billions of years ago.

You might be wondering how a bunch of bonded carbon can really tell us anything about the inside, or history, of our planet. Diamonds are a beautiful and sparkly accessory and if you are fortunate enough to own a top gem-quality diamond, then it will be almost entirely free of such 'impurities'. However, if you want to reveal their true beauty then you'd best find one that is not of the highest gemstone quality, containing so-called inclusions. These are tiny samples of the mantle from which the diamonds originated – liquids and minerals that were encapsulated within the growing crystal at depth, trapped within its cubic carbon structure. We can think of inclusions as a bit like fossils trapped within amber.

While inclusions give a diamond lower clarity, and could also lower its strength, they add great scientific value. By studying the inclusions, scientists can put together a picture of what the surrounding rocks look like at depth, those that created the inclusion, thus allowing them to recreate the birthplace of the diamonds. The majority of diamonds contain these inclusions and, as geologists, we would much rather see a crystal packed with impurities because they offer direct evidence of what is inside the Earth, a little time capsule of information about the origins of the planet.

These are materials that have been held in the diamond crystal structure since it formed, encapsulated within a solid and exceptionally strong covalently bonded jail for, in some cases, over 3.5 billion years. Next, you need to grab a microscope to get a close-up.

Within diamond inclusions is contained a whole host of chemical information about our planet, so they are analysed carefully and painstakingly with scientific instruments, usually some type of highly advanced microscope. Inclusions are commonly used to date diamonds using radioactive decay systems of elements, which can then be used to calculate when the inclusion was trapped within the diamond. We can't use the carbon in the diamond itself because the carbon-dating system only works on younger materials up to around 50,000 years old.

Diamond inclusions have even given scientists proof that the mineral phases they predicted from theory should exist in the lower mantle, really are there. Recently, the rather complicated-sounding calcium silicate perovskite ($CaSiO_3$) was found within a diamond. $CaSiO_3$ is a mineral phase that should only be stable at depths below 650 kilometres (400 miles), at pressures of more than 240,000 times that of Earth's atmosphere at sea level. Because this mineral exists at such depth, we've been unable to sample it before and, obviously, it was not expected that it would exist at the Earth's surface. But the unyielding structure of a diamond was perfect for capturing and preserving the $CaSiO_3$ on its long journey from the deep, allowing us to receive a perfect sample of the lower mantle at the surface and proving that scientists' predictions of what is down there were correct.

And that's not all. Diamond inclusions have also brought to the surface materials that were totally unexpected. One of these is water ice! As unbelievable as it might seem, diamond inclusions have revealed that liquid water happens to exist at great depth within our planet. This water was trapped within growing diamonds as an inclusion that, when transported to the surface, crystallised as a special type of water ice known as ICE-VII. This type of ice is now officially recognised as a mineral. Scientists figured out that these ICE-VII inclusions were trapped as highly pressurised liquid water, only forming ice as the diamond was transported to the surface and subjected to cooler temperatures.

But what is ICE-VII? If we start with what we consider as normal ice, the type we put in our drinks and on which we skate, then that is known as ICE-I. Between ICE-I and ICE-VII there are five other versions of ICE, all forming under slightly different conditions. It's thought that ICE-VII, the kind of ice that is formed when water is compressed at very high pressures, could be present at other planetary bodies such as Enceladus, Europa and Titan, which would suggest it is quite abundant throughout the Solar System, and possibly even beyond. It might not be as unusual as we first thought. However, before it was found in diamonds on Earth, it was thought it couldn't form naturally on, or in, our planet. ICE-VII had been manufactured in a laboratory on Earth, but Earth's interior our planet is known to be too hot for such ice to be stable. Sure enough, this is true. But thanks to the high pressure maintained within the diamond structure, we now have an example of ICE-VII on Earth and, at the same time, have discovered evidence for water deep in the Earth. The water existing at these depths, some 400 to 600 kilometres (250 to 370 miles) down in the transition zone from the upper to the lower mantle, is thought to have come from the surface. It is water that has been transported into the planet in lithosphere dragged down at subduction zones; in other words, lithospheric material that just kept falling. The water in the subducted slab was squeezed out because of the increasing pressures.

The fact that scientists now know for certain that water exists at such depth in Earth has allowed them to re-evaluate how much water our planet holds in its interior. This has important implications for probing into the history of our planet, and in understanding mantle convection and even things like the development of our atmosphere. However, scientists still don't know for sure how much water is down there, or how widespread these pockets of liquid might be. There is still much to learn, but the fact that it's there at all is a key finding.

It is not just the ICE-VII-containing diamonds that are special though. Researchers managed to get their hands on a number of gem-quality 'blue' diamonds that make up just 1 in 200,000 diamonds found. These are rare crystals that are consequently precious thanks to their unique blue tint from the presence of the element boron. The 45-carat, walnut-sized Hope diamond, owned previously by Marie Antoinette, Queen of France in the eighteenth century, is one of the most famous jewels in the world and happens to be a blue diamond. While it had long been known that boron is found within blue diamonds, scientists know that boron is predominantly found at the Earth's surface, or in the shallow crust, so it was a mystery as to how it was contained within diamonds formed deep in the mantle. Even more strange was the fact that other inclusions contained within blue diamonds indicated that these crystals originated in the lower mantle, the deepest regions where diamonds form, and the inclusions provided the first proof that diamonds existed in that region. The question remained as to how the boron got there.

What the group discovered was that the boron probably does come from the Earth's surface after all, and that it rides down into the lower mantle with the material – namely oceanic lithosphere and the minerals contained within it – taken down at subduction zones, just like the water we heard about above. The boron was incorporated into water-rich minerals formed in geochemical reactions between oceanic crust and seawater in a process known as serpentinisation,* but the surprising fact was that these minerals could survive on their journey that deep into the Earth. In support of the previous findings about ICE-VII, there seems to be mounting evidence for a super-deep hydrological cycle in the Earth that we wouldn't know about if it weren't for diamonds.

## The circumstantial evidence

Indirect evidence about the interior of a planetary body comes in many forms. The great thing is that some of this evidence doesn't even require us being on the surface of the planet to obtain it. For example, if we see active volcanoes from orbit around a planetary body, then we know that it is warm and molten inside, whereas extinct volcanoes indicate that the object was once warm, at least at some point in its past. An active, or recently active planetary body, tells us that molten material probably exists at relatively shallow depths so that it can get from there to the surface. But there are other ways that we can glimpse inside a planetary body, some of which don't even require us being able to see the planet itself.

---

* A chemical process whereby a rock, usually of ultramafic composition, is changed with the addition of water into the crystal structure of the minerals. The name derives from the similarity in the appearance and texture of 'serpentinised' rocks to snakeskin.

From afar it is quite easy to detect whether a planetary body has a magnetic field. Earth's magnetic field is powerful, which is a good thing because it helps protect our planet from the solar wind that would otherwise strip away our ozone layer and atmosphere and leave us susceptible to deadly ultraviolet radiation. Our magnetic field exists because of movements in our metallic, molten outer core that together act like a giant, hot dynamo. If the entire core was solid, this wouldn't happen. When we look at Mars, we don't see a magnetic field. What this tells us is that Mars' core today is probably solid, or almost solid, but that doesn't mean it didn't have a liquid core in the past. The presence of large volcanoes on the Martian surface tells us its interior was warm enough at some point in its past to produce them. Mars may just have cooled down too much by the present day, so that its dynamo is now broken.

## Seismic evidence

If we want to quantify the depth, size and composition of the cores of planetary bodies, as well as the existence, or not, of other internal layers, then magnetic fields have done all they can to help. They can only tell us the core is there if it's liquid. Next, I want to discuss the field of seismology, which relies on physics and has the almost magical ability to 'see' inside planetary bodies.

I have always found physics concepts hard to picture, probably because they are often dealing with things we can't really see. I am more comfortable with holding a rock in my hand and being able to make basic assumptions about how it formed from what I can see contained within it. But physics is concerned with particles that are too small to feel, or speeds of light that are impossible to even imagine. This is, partly, why the subject is so fascinating as it can

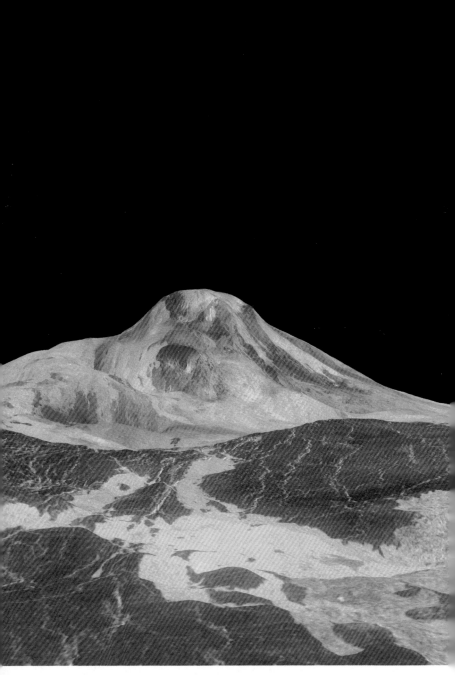

**Above:** Maat Mons, a massive shield volcano and the second highest mountain on Venus, rising eight kilometres (five miles) above the mean planetary radius, in three-dimensional perspective taken by NASA Magellan. Bright lava flows extend to the foreground. Vertical scale has been exaggerated by 22.5 times.

**Left:** The author standing on a peak next to the active lava dome of the Soufrière Hills volcano on Montserrat in 2008. Steam can be seen emanating from the rock that makes up the freshly extruded lava dome.

**Right:** Destruction of a building in Plymouth, the capital of Monserrat, by pyroclastic flows and lahars that began in 1995.

**Below:** The Soufrière Hills volcano on Montserrat in 2008, showing the active lava dome producing steam and a plume of gases.

**Above:** The Hawaiian archipelago in the Pacific, as seen from space.

**Below:** One of the lava flows emanating from the Kilauea volcano during the 2018 eruption.

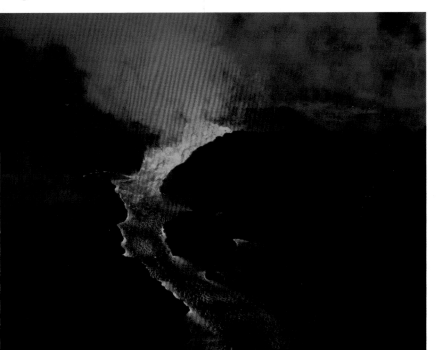

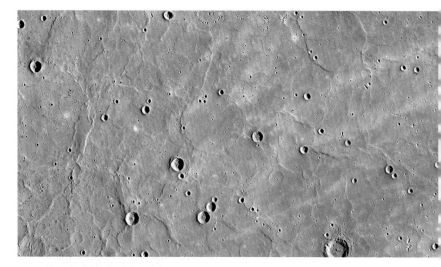

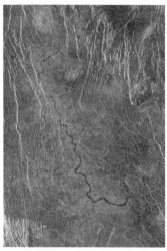

**Above:** Expansive smooth plains on Mercury's northern region, created by volcanic lava flows that cover older craters, known as ghost craters.

**Left:** Radar image taken by NASA Magellan showing a 200-kilometre (124-mile) segment of sinuous channel on Venus. These channel-like features are common on the plains of Venus and, whilst they look like rivers on Earth, they are thought to have been formed by lava eroding a path over the plains as it flowed.

**Below:** A view of volcanic plains on Triton derived from topographic maps acquired by NASA Voyager in 1989. The cantaloupe terrain is clearly visible.

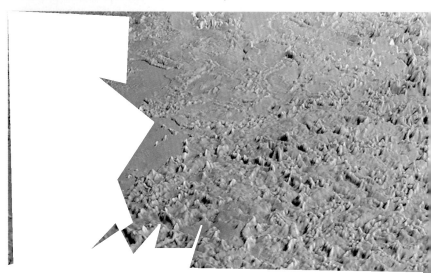

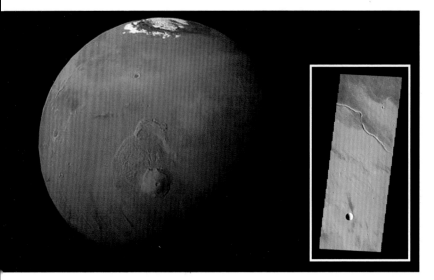

**Above:** Orbital view of Olympus Mons on Mars, the tallest volcano in the Solar System, standing at 22 kilometres (13.6 miles) high. **Inset:** Tharsis volcanic flows on Mars showing a channel that was most likely carved by the flow of molten lava.

**Right:** Plumes on Enceladus made up of jets of icy particles.

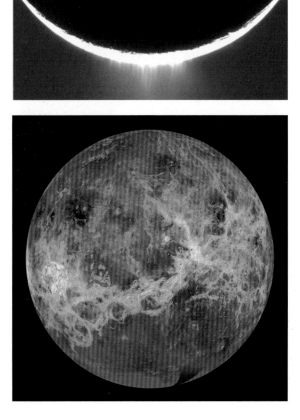

**Right:** Global image of Venus captured by NASA Magellan spacecraft.

**Left:** Normally enveloped in a thick atmosphere, the NASA Cassini mission used its Visual and Infrared Mapping Spectrometer (VIMS) to see through Titan's haze to the reveal the surface. The colours seen reflect the different compositions of material.

**Right:** A NASA Cassini spacecraft radar image of Titan's north polar region showing seas and lakes (in blue) made of liquid hydrocarbons.

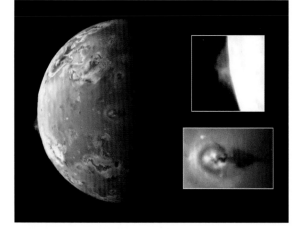

**Left:** Colour images acquired by the NASA Galileo mission showing volcanic plumes on Io.

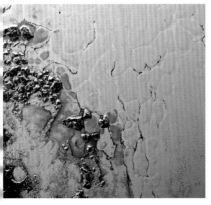

**Above:** Ice convection on Pluto's Sputnik Planum showing cells of ice. The image is about 400 kilometres (250 miles) across.

**Above:** Enhanced colour image of Pluto (lower right) and one of its moons Charon (upper left).

**Left:** An enhanced colour image of Ceres showing the distinct textures and compositions. In the centre of this image is the brightest region, Cerealia Facula, in the Occator Crater. Bright material is generally thought to be salt-rich, excavated from Ceres' crust.

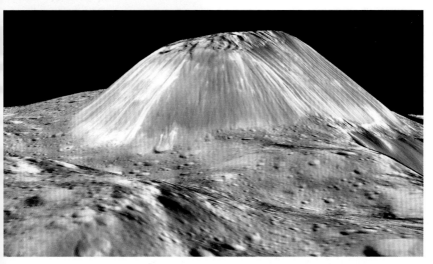

**Above:** Ahuna Mons on Ceres rises 4000 metres (2.49 miles) tall and the bright streaks that run down its flanks are salt deposits from saltwater and mud that rose from within Ceres.

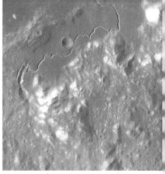

**Above:** Mare Serenitatis is one of the lunar mare and situated close to Mare Imbrium. Its diameter is 674 kilometres (419 miles).

**Below:** This image of the near side of the Moon shows clearly the lighter patches of the lunar Highlands and the darker patches of the lunar mare.

**Above:** Hadley Rille is a meandering channel carved on the Moon by a lava flow around 3 billion years ago. It is more then 120 kilometres long, up to 1,500 metres wide and more than 300 metres deep in places.

explain things most people didn't even consider needed explaining. But sometimes it takes a bit of faith for us non-physicists to accept what we're being told. Nevertheless, physics within geology is an incredibly useful field of study and one that I have embraced. Geophysicists can access the parts of the Earth that geologists fail to reach – they allow us to peek into the true depths of a planet, all the way through the core and out the other side.

In relation to volcanoes, geophysics makes up one part of the scientific evidence that can be used to understand when a volcano will erupt, and what is going on stealthily beneath its apparently cool and quiet exterior. This form of geophysics relies on measuring seismic activity, or earthquakes, associated with magma moving below a volcano and cracking apart the crust it sits within. Using seismographs to measure earthquakes associated with volcanic activity allows scientists to figure out if magma is moving and at what depth, and if there's a chance that the system will be active at the surface soon. It was around the 1980s that seismic data first started to be used to predict volcanic eruptions. Although there were a few smaller successes throughout the 1980s and 1990s, in 2000 an eruption of Popocatépetl was successfully predicted 48 hours in advance by the National Center for Prevention of Disasters in Mexico City. The forecast led to the evacuation of tens of thousands of people from the region. The resulting eruption was the volcano's largest for a thousand years, yet no one was hurt.

However, often the seismic signs of unrest are coupled with other factors such as measurements, observations of gas release and deformation of the ground around a volcano as the magma pushes up to the surface. This provides the clearest picture as to whether an eruption is imminent and prevents the rare but costly mistake of scientists announcing that an eruption is due – and suggesting people evacuate the area – when there is no eruption occurring, as we saw in Chapter 2.

But by measuring these same earthquakes – along with additional seismic events triggered by other fault systems and volcanoes around the world – scientists don't just get to find out more about volcanoes; they can actually peer deeper into the Earth. Earthquakes let scientists map out what the interior of our planet looks like and even what it is made from. I previously likened this to medicine: how a radiographer uses X-rays to image the inside of a person's body. X-rays can see tiny breaks in a bone because they travel differently through tissue of different density, and thus an image is created showing the results. In the same way, earthquakes allow us to image the internal layers of the Earth because the seismic waves they produce speed up or slow down depending on what they travel through.

When an earthquake is triggered, it doesn't just affect the region directly around it. When a rock is broken or moved underground – the incident that initiates the earthquake – energy is released as seismic waves that emanate out in all directions from the hypocentre, the focal point of energy release. The epicentre is the spot on the surface above the hypocentre. Once these waves travel through and around the globe, they can be detected anywhere and measured in the places scientists have placed a seismometer. These waves come in two basic varieties: body waves and surface waves. This is useful because they have different properties and further variants that allow us to deduce information about what they have passed through. We need to understand a little more about these waves if we are to comprehend how they are helpful.

**Body waves** can be either Primary (P) or Secondary (S). The P waves can be thought of as pushing out from the hypocentre and they act much like a compressional sound wave, shoving and pulling the material through which they pass. They can travel through liquids and solids and, being the fastest waves, are the first to be detected at a seismic station. The S waves, on the other hand, are also known as shear waves. They are slower than P waves and can only travel through solid material, since it is not possible to apply a shear force to a liquid or gas.[*]

**Surface waves** come in two varieties: Love and Raleigh. The former produces a side-to-side motion and the latter a rolling action. They are slower than body waves, and are unable to penetrate through the planet; instead they pass around it. As such, they are usually most responsible for the damage and destruction at the surface.

The different properties of these seismic waves mean that they are slowed down differently, or stopped completely, depending on what material they travel through inside the planet. Let's picture the situation. An earthquake occurs on the subducting Pacific plate beneath Japan. While the surface waves are potentially causing colossal damage in Japan, the P and S waves were already sent out through the Earth and seismic stations around the globe have begun recording their arrival, possibly at the same time as residents are experiencing ground shaking, depending how large the earthquake was. The data related to these waves is collected by seismographs at seismic stations around the world and shared so that the seismic

---

[*] Shear forces occur when unaligned forces push one part of a solid object in one direction and another part of it in another direction, like when a deck of cards is pushed one way on the top and another on the bottom, causing the cards to slide over one another.

network can calculate very accurately where the earthquake originated, and just how large it was in terms of how much energy it released. But at the same time, scientists are using this information to find out about what's inside our planet. So how does this work?

Andrija Mohorovičič was a Croatian meteorologist who became a seismologist when he discovered a few interesting features of seismic waves. In 1909, Mohorovičič was studying data collected by seismographs placed prior to an earthquake in Zagreb, Croatia and he found that some seismic waves seemed to travel faster than others, but also that the waves apparently didn't travel in straight lines. As you might recall from having a basic understanding of school-level physics, waves change direction when they pass through materials of different density. This is known as refraction and you will probably know about it, but perhaps not necessarily by name, because this is what happens when a straight stick looks kinked when half-immersed in water. The same sort of thing happens to seismic waves as they travel through the Earth: they follow curved paths because the density of the planet increases with depth. The result is that the waves are sped up as they progress to the core. Mohorovičič used this information to conclude that the Earth's crust was less dense than the material below it, because the waves travelled slower at the surface. In fact, his name was given to the major so-called 'discontinuity' between the base of the crust and the top of the mantle, sometimes shortened to 'Moho' where there is a sharp change in density.

But some waves also get reflected off different boundaries in the Earth. In the early 1900s, the German-American geologist Beno Gutenberg used this observation to conclude that the outer core must be made of liquid because the S waves were unable to penetrate through it. In fact, some seismometers won't pick up the arrival of any S waves, just because the earthquake that sent them out might have been on the other side of the planet and the shear waves were unable to travel through the liquid outer core, thus 'disappearing' and leaving an apparent seismic shadow on the opposite side of the planet to the location of the earthquake. The seismic shadow is the region on the Earth's surface where S waves are not detected after they have radiated out from an epicentre. By contrast, those shear waves that take a different route, avoiding the liquid outer core, can be detected on the opposite surface from the epicentre after passing all the way through the Earth.

Not only did Gutenberg figure out that there was a sudden density contrast between the liquid outer iron core and the silicate mantle (which is now termed the Gutenberg discontinuity), but he also found the inner core must be at great pressure, forming a dense solid through which P waves can travel very quickly. In this way, seismic waves were producing a detailed 'X-ray' image of the parts of our planet that we've never been able to access.

The great thing about earthquakes is that they don't just occur on Earth. Well, technically they do because of the name, but other planetary bodies experience quakes; they could be called marsquakes, venusquakes or moonquakes, depending on where they occur. While Earth is the only place where we've been able to place an extensive grid of seismic stations to provide a clear picture of the total seismicity of the planet, it isn't the only place in the Solar System to have them. Astronauts deployed four seismometers on the Moon during the Apollo landings and there is even one on Mars, placed by the NASA InSight mission in 2018.

While these instruments don't form as extensive a seismic detection system as we have on Earth, they still inform scientists about the interiors of these planetary bodies, including if they are still active. The lunar seismometers were switched off in 1977, but the quakes they detected prior to this were a surprise to scientists because they didn't know if the Moon would still be active. While moonquakes are not caused by plate tectonic movements like the quakes on Earth, instead probably being the result of old faults and cracks relaxing as the Moon continues to cool, their occurrence still allowed scientists to calculate things about the inside of the Moon. They found it hosts the same type of layering as our planet: a solid core and a small, partially liquid outer core, a mantle and a crust. However, because the original moonquake data were said to be quite 'noisy', meaning there was interference on it that needed to be removed, it actually took until 2011 for scientists to have enough computing power to develop techniques to clean it up adequately to make these conclusions, by picking apart the detail in the seismic data.

Mars only has one seismometer, called SEIS (Seismic Instrument for Interior Structure). After touchdown, SEIS was carefully moved from the InSight lander to the ground before being covered with a protective coat of chainmail to prevent it from being knocked around by Mars' windy weather. Since then it has been detecting marsquakes to build up a seismic picture of the red planet. Just like on Earth, these data, along with temperature information gained from a probe that was set to drill down into the Martian crust,* will allow scientists to understand more about Mars' current internal state.

### Evidence from the skies

Without the occurrence of seismic events combined with such clever geophysics, we would struggle to know the

---

* Although at the time of writing it appears that the 'mole' was unable to penetrate to the required depth for these measurements to be made.

exact depth of the internal layers of the planets. But while this technique helps us learn about the structure of the planets, it can't tell us a huge amount about the composition of the materials that make up those layers, apart from information about their relative densities and states. Geophysics needs to be combined with other indirect observations for us to learn more, and one of these techniques involves looking at asteroids. Asteroids are the small, rocky objects that make up some of our Solar System. They can be thought of as the building blocks of the planets, as they contain all the same ingredients, having formed in the same location and at the same time as the inner planets. The asteroids just never grew large enough to become fully fledged planets and so they gathered together in the asteroid belt located between the orbits of Mars and Jupiter.

Asteroids have experienced a lot of collisions with other asteroids in their past and this sometimes caused them to break up into smaller pieces that orbited around the inner Solar System for many years. These little pieces of rock sometimes ended up colliding with a planet, and continue to do so, and when they land on the surface of another planet they become a meteorite. Meteorites – pieces of space objects on Earth – can be composed of rock, but some are pure metal, made up of iron alloyed with nickel and very small amounts of other precious metals. These hardy pieces of asteroids transported to Earth inform us about the very deepest parts of our planet, and the bodies that surround us.

Metal meteorites are almost certainly a sample of the core of an asteroid that had once grown large enough to melt and gravitationally segregate into a core and a mantle, just like the terrestrial planets, before being smashed to pieces in a collision with another space rock. When the asteroid was demolished, pieces of it ended up scattered across the Solar System, and those that fall to Earth can be used to study not only the processes that formed the asteroid, but also those that formed the planets in general. Metal meteorites can be a direct sample of an asteroid core, and an indirect sample of Earth's core. We know this because the density and composition of such rocks is the same as we expect for the material in our core based on geophysical predictions.

That's not where the scientific beauty of meteorites ends, though. Other meteorites, those that are rockier, can tell us about the overall composition of our planet, allowing scientists to glimpse the beginning of Earth as a planet, before it had internally differentiated into its current structure of core, mantle and crust. These so-called chondritic meteorites (containing small mineral granules called chondrules) are special because they did not differentiate into a core and mantle like the metal meteorites. Instead, they are composed of the earliest materials available in the Solar System, which combined from the cloud of gas and dust originally surrounding our young Sun and from which all the planetary bodies formed. The chondritic meteorites have remained almost unchanged since they formed, preserving their original composition.

Since the Sun comprises over 99.9 per cent of the mass of the Solar System, we can be quite certain that Earth's overall, or bulk, composition is similar to that of the Sun, since it formed from that same disk of material that surrounded the Sun at its birth. The bulk composition of the Earth is, therefore, also similar to the chondrites. By studying these meteorites, scientists can backtrack our planet to its primordial, or birth, composition. Plus, by knowing roughly the composition of our crust and mantle at the present day – because we've been able to sample those – we can figure out what the missing parts are. This is mainly the stuff we can't see because it is locked away deep in the core. We know it must contain heavy elements to balance out the light crust, and from what we have seen is contained within the chondrites. We also know from seismic data that the material in the core is very dense, so metal meteorites fit the bill. However, iron and nickel alone are too dense to account for the core on their own, and so scientists think that it must contain about 10 per cent of a light element too. The identity of this mystery element remains unknown for now, but something like oxygen is a strong contender, despite the problem of understanding how it would have got there and not remained nearer the surface.

By knowing this information about our own planet, we can apply it to the rocky planetary bodies surrounding us, as they all formed in a similar part of the Solar System to Earth – relatively close to the Sun – from roughly the same chondritic ingredients. As such, we can say that the cores of Mercury, Venus and Mars – although different in size from our own – are very similar in composition. In fact, as we now know, sometimes by studying one planet we can learn a great deal about others, without even having to go there.

## Scratching the surface to reveal the truth

As you can see, by combining information about a planetary body's density, seismic activity, magnetic field and so on, we can make some reasonably good assumptions, even conclusions, about what makes up their interiors. Furthermore, while looking at the surfaces of planetary bodies tells us a great deal about their internal dynamics it can, at the same time, allow us to learn their history. Craters, for example, can tell us how many times the body was impacted by objects from space, either comets or asteroids. The more craters that are detected, the older that surface is assumed to be, because cratering is a cumulative process. If no, or hardly any, craters are seen then we can make some good assumptions that the surface is 'fresh', that it is a geologically active planet or has been in the recent past. Earth is one of those places thanks to its dynamic and hot interior, plate tectonics, and the geological activity that continually resurfaces its exterior, destroying the ancient evidence of impacts from space. The Moon, on the other hand, is heavily cratered, suggesting that the surface we see today has been there a long time. In fact, the *youngest* rocks on the Moon could be around 1.5 billion years old. These values are estimates based on crater counting, but meteorite NWA 773, a lunar meteorite that landed in north-west Africa, has an age of 2.8 billion years, which is a directly measured value. Of course, we haven't sampled the entire Moon, so it could be that there are younger rocks that haven't been sampled and analysed yet.

It's only within the last handful of decades that we've departed from Earth with spacecraft to venture out and visit other planets and moons within our Solar System, so we haven't really had it within our psyche to think of these alien worlds as active in the same way as Earth. The Jovian moon of Europa is one of the places we are most interested in. In the late 1990s, NASA's Galileo spacecraft discovered Europa's geologically young surface, which is made of ice, and in 2018 researchers carried out a re-evaluation of the data obtained to find that Europa displays water-plume activity. Coupled with the fact that images of the moon reveal the smoothest surface of any planetary body and a very high albedo (high light reflectivity, thanks to the ice), this indicates a geologically fresh and young surface. In fact, Europa's exterior is estimated to be just 20 to 180 million years old. It is probably thanks to the plumes, or cryovolcanoes, that resurface Europa's icy exterior, that the surface is so new. But Europa's smooth skin also hints that its interior must be dynamic and active too.

What I'm really stressing here is the importance of studying the outer shell of a planetary body – its crust, whether made of rock or ice – if we want to learn more about its history and about the internal parts we can't access very easily, if at all. In fact, when space missions visit other planetary bodies, usually the simplest and cheapest exploration they do is to study the surfaces by taking pictures. For this we just require a set of cameras that are good at taking pictures from afar, as a spacecraft orbits or flies past a planetary body.

Of course, planetary bodies are probably best explored by humans, and if not humans then robots or rovers. But to get to the stage where we can send humans or robots, we need to learn a little about the places they will visit. Where exactly will we place the humans to explore the planet? In order to check they will touch down on a safe yet interesting region we must first obtain images of the location in question. Luckily, all planetary bodies can be studied in detail remotely, either from Earth itself or from spacecraft in orbit. Furthermore, landing robots, or even humans, on the surface of planets to drill into and sample them is technically challenging and expensive, hence it is not a regular occurrence. Of course, when it does happen, it provides us with first-class data about the planetary body. Nevertheless, the simplest first steps in planetary exploration involve imaging.

We can take the NASA New Horizons mission to Pluto and its surrounding moons as an example. Before the New Horizons encounter with Pluto, the best images we had of this little world were taken by telescopes in the years

following Pluto's discovery in 1930. They showed a
planetary body that was definitely spherical, but one that
could only be described as blurry thanks to the very low
resolution of the images. While the scientists were correct
in thinking Pluto's surface was composed of ice, they were
unable to make out any detail and all they could conclude
from the images was that it contained some general swathes
of intriguingly different coloured regions. This wasn't
very informative. New Horizons was launched to get up
close to Pluto. Despite it being the ninth planet in the
Solar System at the time of launch in 2006, before being
demoted the same year to dwarf planet, we still knew so
little about Pluto. The theory was that the planet was
probably just cold and dead, thanks to its vast distance
from the Sun and small size, being just two-thirds the size
of our moon.

The images returned by the New Horizons spacecraft
helped to cement Pluto's place in the Solar System hall of
fame. Already a bit of an oddball, Pluto orbits the Sun on a
different plane from all the other planets, also differing
from the other small icy objects – most of which are
comets – in the Kuiper Belt. The images beamed back to
Earth revealed an object that indeed was composed of ice,
but that hosted intriguing and somewhat unexpected
geological features: dunes made of methane ice, glaciers of
nitrogen flowing down slopes, and a lack of craters. There
is no solid rock on the surface of the dwarf planet, yet it
appears to have features just like our own rocky planet
because its 'bedrock' of ice acts in a similar way to rock on
Earth. Parts of Pluto's surface are clearly 'fresh', lacking the
tell-tale pockmarks of cometary impacts that would be
expected if activity – in the form of cryovolcanoes and
other tectonic activity – hadn't resurfaced it. In fact, the
smooth and craterless portions of Pluto's surface allowed

scientists to quickly conclude that the surface of this tiny world was potentially less than 10 million years old – an infant on geological timescales.

Suddenly scientists were realising that this now dwarf planet was active in the geologically recent past and possibly even at the present day, as it could resurface its outer shell. This was completely unexpected. Pluto is a small, cold ice ball sitting very far from the Sun. As such, it was thought to be inactive, much like the icy comets that surround it and which preserve materials, unchanged, from the formation of the Solar System some 4.6 billion years ago. The key thing is that Pluto has been active recently, and this means that it must still be warm inside in order to fuel its surface activity. But scientists just didn't think this was possible; it really should have lost any heat it may have had from formation and cooled completely by now. Nevertheless, the data coming back from New Horizons is revealing new insights. This icy world is fascinating and is one of the places that has forced us to re-think how we view geology; it's not all about rock, after all.

How, and at what speed, a planetary body cools are features we'll learn more about in the following chapters, because every object has a different story to tell. Pluto will necessarily feature in this discussion. While it is true that small bodies lose their heat quicker than large bodies, and those further from the Sun receive less of its heat and so cool quicker too, there are other ways that planetary bodies can find and retain heat. Pluto has obviously found a way to stay warm despite its distance from the Sun, in order to fuel activity at its surface, but the way it does this needs some exploration. The important thing to note here is that by taking a fleeting, fly-by glimpse at the surface, scientists were able to deduce a great deal about the insides of this fascinating world, and of the moons that orbit it, and are

continuing to discover more as the data is analysed in
further detail.

## Spouting evidence

A final piece of evidence we should discuss is geysers, partly
because they are an interesting volcano-related phenomenon
on Earth, but also because similar features seem to have
been observed in space. Some of the most famous locations
on Earth to go geyser-watching are Yellowstone National
Park in the United States, which is home to over half the
world's geysers, but also Iceland, New Zealand and the
Kamchatka Peninsula in Russia, to name just a few. We
can thank Icelanders for the name, with the word 'geyser'
deriving from 'geysir' in Iceland, the place famed for its
episodic discharge of hot spring water.

On Earth, geysers discharge water that has come from the
surface as rain or snow, having filtered down into the
bedrock. It is known as meteoric water and there is nothing
volcanic about the water itself. However, the reason that
water at geysers is forcefully ejected onto the surface is *because*
of volcanic activity nearby. Geysers have magma sitting
beneath them and it is the magma's heat that warms up the
meteoric waters, eventually heating them to boiling point so
that they can't help but escape from the ground as steam and
liquid water. These meteoric, or hydrothermal, waters are
just one of the important fluids associated with volcanoes and
eruptions. Geysers on Earth erupt water and volcanoes erupt
molten rock, but both liquids are associated with the volcano
and simply represent places where relatively hot material
makes its way energetically from the interior to the surface.

Geysers are fairly uncommon on Earth because they
require a few conditions to be met, such as abundant water

recharge and magmatism, large cracks for the water to get into the crust and cavities to store it. With the increasingly modern use of water for energy to meet growing population demands in many regions, some geyser fields are now extinct, their supply of meteoric water having been diverted. While steps have now been taken to preserve Geyser Valley in Kamchatka, some similar features in New Zealand have disappeared as water has been redirected to electricity power stations.

Geysers are incredibly important in helping us to understand the dynamics of volcanic eruptions in general. As we've seen, while geyser eruptions on Earth don't tap magma itself, they can still be used to understand what goes on underground, including the processes that eventually force liquids and gases out of the crust and into the atmosphere. Most geysers erupt more frequently than volcanoes, thus providing a wealth of data about the physics and dynamics of eruptions more rapidly than their rocky counterparts. Old Faithful geyser in Yellowstone National Park, for example, is highly predictable, erupting every 44 to 125 minutes since the year 2000. This has allowed scientists to set up equipment to understand the eruptions, knowing that they will capture a lot of data.

When it comes to geysers on other planetary bodies, we didn't know any existed until the first plumes of material were captured emanating from Triton, one of Neptune's moons, in images beamed back by the Voyager spacecraft in the late 1980s. It is not yet fully understood what powers Triton's geysers. One idea suggests they are a result of the, albeit weak, sunlight heating Triton's surface. The sunlight received by Triton could be just enough to heat its dark organic material present below the translucent nitrogen ice caps, and which efficiently absorbs the solar energy to warm the adjacent ice – like a solid greenhouse effect. Researchers suggest that, while this level of heating might increase the temperature of the nitrogen ice by only 4°C (7.2°F), it could be sufficient to produce gas bubbles, which collect together, becoming over-pressurised before they erupt out at high velocity, producing the impressive high plumes observed by Voyager. Another theory suggests these geysers could be created from internal heat within Triton.

The first idea, that they are a result of sunlight, would mean they're not volcanic in nature, and are much more akin to Earth's geysers. After all, geysers on Earth are not themselves volcanic, as they are not erupting molten rock. The second idea would suggest they are properly volcanic in nature, being powered by internal heat transfer within the interior of Triton and erupting Triton's own brand of molten bedrock. We'll learn more about Triton and its activity in future chapters; it is intriguing as an active world so far from the Sun.

Many decades after the Triton geyser discovery, we still don't think geysers are particularly common in space. Yet very similar-looking features have since been found on moons of some of the planets, including Europa and Enceladus. As we've seen elsewhere in the book, the activity on Enceladus was first spotted by the Saturn-orbiting NASA Cassini spacecraft in 2005. With its solid surface of ice and salty, liquid ocean below, Enceladus is an interesting place because this far from the Sun, we expect everything to be frozen solid. But scientists have discovered that Enceladus' subsurface ocean is kept warm by the influence of the immense and powerful Saturn. The gravitational forces of this giant planet twist and pull the 530-kilometre (330-mile)-wide moon as it goes about its elliptical orbit, squeezing and stretching its insides. These forces create frictional movements within Enceladus, grinding its insides together and generating internal heat. This is a process known as tidal flexing, or tidal heating. It is enough to keep the subsurface ocean above Enceladus' core warm enough to remain liquid. The icy exterior of this planetary body truly hides an excitingly active world below. The heat within the subsurface ocean provides the energy to fuel the plume eruptions from within it. These are a type of cryovolcano, releasing fluids from Enceladus' salty ocean through cracks in the ice that are periodically opened up and closed again because of the tidal forces. Enceladus' plumes, therefore, are not like Earth's geysers as they are themselves volcanic. The material in these plumes not only falls back as snow onto the surface of Enceladus, but also becomes the source for one of Saturn's rings, known as the E-ring.

Whether we are looking at geysers or volcanic plumes streaming off the surface of other worlds into space isn't actually important because they both indicate that that place is active. At the same time, the eruption of plumes of vapour and liquid allow us to glimpse inside these planetary bodies, places where we've not had the chance to land a spacecraft yet. Whereas on Earth our geysers are firing out surface waters that have filtered back into the planet, on other planetary bodies the plumes are transporting subsurface liquids to the surface, from potentially great depths. By detecting the molecules being ejected in their plumes, we can discover some of the materials that make up their interior.

Active ocean worlds are important because a dry, barren, inactive planet is seriously unlikely to host the necessary conditions for life. As such, planetary bodies like Enceladus, and similar ones like Europa, become important locations to concentrate our search for life elsewhere in the Solar System. Enceladus meets many of the requirements that scientists think are necessary for life. The fact that hydrogen has been detected – while very abundant in the Solar System – is especially exciting as its production on Enceladus is thought to be related to hydrothermal reactions between its salty ocean and its hot rocky core. As seen in Chapter 5, hydrogen at mid-ocean vents provides a source of chemical energy that microbes use as their food, with the warmth from the vent helping fuel the process of chemosynthesis. If such a process occurs at the base of Enceladus' ocean, then it represents a way for potential life to have the necessary fuel to thrive and shows that the organic molecules detected by Cassini are biological in origin.

For now, though, that is a bit of an intellectual leap to make because we are still unsure how life itself originates. It's all well and good having the right conditions for life to thrive, but it has to have found a way to develop there from the start. Unfortunately, Cassini came to the end of its mission before answering many of these questions, so we will have to wait patiently for more information from future missions if we are to confirm whether or not we are alone in the Solar System. However, without Cassini having detected those plumes, we'd have no idea what was down there. Now we know it's worth investigating Enceladus further.

# CHAPTER SEVEN

# Warming Up

Whether we're looking at the stiflingly hot volcanoes of Earth, or the freezing cold cryovolcanoes of Enceladus or even Pluto, they have one thing in common. The important similarity that makes them the volcanically active places they are – whether hot or cold – is a difference in temperature between their insides and outsides. It might not sound very important, but it is this contrast that fuels their activity. It really all comes down to a nifty bit of physics.

Put simply: heat moves. Heat has a natural tendency to flow from a higher temperature reservoir to a lower temperature one. In planetary bodies, the result of this heat transfer causes materials to become buoyant and move in the form of ice or rock, whatever is their 'magma', so that they rise to the surface of a body. What erupts at the surface simply depends on where the planetary body is located within the Solar System, where it formed, and from what material it was originally made. Whatever the case, as soon as a planet or moon forms, it is looking to cool off, to lose its initial heat and solidify. Nevertheless, there are some factors working against this cooling.

The way in which planetary bodies produced their initial heat, even if they have always been formed of ice, and how they cooled down after, is not necessarily a shared process; each might have got its heat from a different source, such as the heat of formation (primordial heat) or tidal heating, for example, and they might have cooled in different ways: one might have been more volcanically active, another might have had a surface layer that insulated heat. Each planetary object heated up and cooled down in a different way, and the mix of processes is unique to each one.

A planetary body's cooling history is not simply related to its location in the Solar System, as there are other factors at play. Just because a planet is close to the Sun doesn't mean that its interior is necessarily hot, and those far from the Sun are not necessarily cold. It turns out that some planetary bodies are self-sufficient, producing their own heat, while others rely on generating it from the effects of external forces acting on them from neighbouring planetary bodies. I'll explain more about this shortly. Yet other planetary bodies are just geologically dead, no longer able to generate heat for themselves or find a way to get it from elsewhere, having cooled very quickly after they formed.

The amount of heat a planetary body can trap or generate itself, and how quickly it loses it, is an important control on the overall evolution of the body on a very grand scale. A planetary body that has cooled too quickly or slowly has no chance of hosting life. Without heat, a body itself cannot be active, and without activity moving heat around from the inside to the outside, surface processes such as geological activity, the growth of an atmosphere and the maintenance of liquid water or other solvents at the surface are not possible. In turn, to have active biology on the surface or subsurface of a planetary body requires that the body is cool enough to allow for sustained liquid water on its surface.

## A planet-sized hot-water bottle

As we've seen, volcanoes themselves – whether made of fire or ice – are one of the expressions of a planetary body releasing its pent-up heat. At the same time, when volcanoes are still active, they are one of the most obvious signs that a body still has internal warmth and is, therefore, not geologically dead. Even if we have no idea of the temperatures within a planetary body, if we see active volcanoes, we know it still has a significant amount of stored heat. Because of this, as we've seen, when we view a planetary body from afar, we can tell a lot about its insides just by observing its volcanoes, whether they are active or not.

Once a planetary body cools past a certain point, it is no longer able to power its volcanoes because it can't turn solid material into liquid. This point is reached long before it is frozen completely solid and immobile inside. Once volcanoes are no longer active, any remaining warmth inside the body will continue to dissipate out into space, via conduction through its interior. Conduction is, in fact, the main way that the Moon and Mars are continuing to cool at the present day, as they have passed the point where their insides are warm enough to convect molten rock, despite this being a feature of their history.

At the present day our own planet is kept perfectly warm, maintaining a surface of liquid water and a climate that is neither too hot nor too cold for life to thrive. Earth is currently experiencing a steady state in its temperature; we're not talking about climate here, but rather the temperature of the body itself. This means that our planet is making as much heat as it's losing from its core. Yes, that's correct: the Earth creates its own heat, just like a boiler. Well, it's actually a bit more like a mix of a storage heater linked to a nuclear reactor, but we'll come on to that shortly.

The heat made by the Earth has to go somewhere and, of course, this is what fuels its many volcanoes thanks to the plate tectonic movements brought about by its warm, convecting mantle. Most of the heat inside the Earth is transferred by the motion of matter. We saw earlier how the planet's solid innards convect and move around at a geologically slow pace, and the tectonic plates sitting above take part in these motions, moving entire continents over geological time.

But the question of why Earth is still attempting to cool off, when places like the Moon are seemingly cold and dead, is a complex one to answer. The heat that drives the large-scale processes on our own planet, and which helps to provide the pleasant surface conditions we experience, is the result of different sources. You might think that our comparatively close distance to the Sun helps to keep our planet cosy and habitable. Here you'd be partly correct, as the Sun's energy certainly helps to drive surface processes such as ocean circulation, atmospheric processes and biological activity. However, it is unable to penetrate deep into the planet. The interior of Earth, particularly the upper crustal layers, is dominated by silicate minerals and these are particularly poor thermal conductors. As such, the heat from the Sun penetrates only about 20 metres (65 feet) into the planet, just a snip of the 6,371 kilometres (3,959 miles) distance to the centre of the Earth. Yet, as we've seen, scientists estimate that the Earth's core sits at around 5,000–7,000°C (9,032–12,632°F), which, as we've already learnt, is roughly the same temperature as the surface of the Sun.

In general, the amount of solar energy received by planetary bodies is too weak, even on the surface of Mercury, the closest planet to the Sun, to drive volcanic activity. However, as we saw in the previous chapter, there is one object in the Solar System that might buck the trend and it is Neptune's moon, Triton. Triton may require very little solar heating to tip it over the edge and produce activity at the surface. This example shows how we don't necessarily require extremes of heat for activity to take place, particularly if we're looking at the icy planetary objects. Of course, if we want to melt and move rock around, and erupt it explosively, then we certainly require some larger relative differences in temperature, something that the terrestrial planets have managed throughout their history.

There is certainly no shortage of heat at depth in our own planet, but if it doesn't come from the Sun then we need to question where it does come from. About half of the heat flowing from the planet at the present day has, in fact, been here for the entire 4.5 billion years since it formed. Earth has never managed to cool down after the fraught and violent formation of the Solar System. This is Earth's primordial heat and it is literally the kinetic heat it accumulated when it was growing into a planet: its first heat. Much of this is stored in the core, and it has been able to hold on to it there so effectively because it is surrounded by a huge chunk of mantle made up of silicate minerals. Remember, those silicate minerals that prevent the Sun's rays from penetrating deep into the planet also act as an efficient insulator to keep Earth's primordial heat locked up.

## Bang, smash

We're not the only planet to benefit from primordial heat. All the planets began life warmer than they are today thanks to the energy released during their violent process of conception. Some of them have just managed to hold on to their early thermal energy better than others.

All the objects within the Solar System – planets, moons, asteroids and comets – were created from the material contained within the protoplanetary disk of gas, dust and ice surrounding our young Sun soon after its birth. The process of planetary accretion describes how the various objects in the Solar System combined their ingredients from this disk by gravitational attraction, which allowed them to grow into the large bodies they became all those billions of years ago. But while planetary formation was relatively speedy on geological timescales – taking around 100 million years – the process was far from serene, which has important implications for understanding where the heat comes from.

As the Solar System's major objects grew larger and larger, they started to experience heating episodes brought about by the physical and violent processes they underwent during their formation, but also simply because of their growing size. Just before the final eight planets of the Solar System were fully established, there were many more slightly smaller objects called planetesimals – the precursors of the planets. Once these bodies reached roughly a kilometre in size, they were able to attract each other through mutual gravity and thus combine and grow larger.

Needless to say, the coming together of these objects was far from peaceful. Planetesimals and smaller planetary bodies were chaotically hurtling around the infant Sun as they hadn't yet managed to settle into the smooth and clean orbits inhabited by the planets today. Collisions between these objects were, therefore, frequent, and when they occurred, they were violent and could be completely calamitous in some cases. If you imagine smashing two planetary-sized bodies together at tens of kilometres per second, it is easy to understand how their temperatures would rise, with the kinetic energy of the impact being converted to thermal energy on collision.

Of course, when planetesimals smashed together they might be shattered to smithereens. But if they impacted at just the right speed and angle, they could amalgamate to form even larger bodies thousands of kilometres in diameter. It is estimated that such accretional energy gave rise to temperatures of around 10,000°C (18,032°F) within planetary bodies at this stage of their growth, although this value is somewhat dependent on the size of the bodies involved and the speed and angle with which they impacted each other. Some of the impact heat was re-radiated straight back into space, but a lot of it was retained in the resulting body, locked deep in its interior. In many cases, the heat generated during the impacts could turn a once rocky planetesimal into a molten one.

The objects that survived these violent processes were the eventual winners of the Solar System formation game, the ones that went on to form the planets. The planets were large enough to gravitationally clear a clean orbit around the Sun, flinging away or engulfing any smaller remnant objects that lay in their path, those that had failed to be collected by a large planetary body already. This would include some of the asteroids, which can, in many ways, be viewed as failed planets.

You'd think that when this turbulent period of Solar System construction had ceased, the planets, remnant asteroids and comets could settle down and begin cooling off from their frenzied start. However, they didn't do so straight away. Although the planets had established their orbits around the Sun, meaning that impacts were much less frequent – particularly large ones – several processes occurred that acted to heat them up once again.

By this stage, many of the planetary bodies had grown so large that they managed to generate their own heat simply because of their immense bulk. As more and more material accreted onto a planet, the matter at the centre was increasingly squashed, thus experiencing a higher gravitational load from the material above. Higher temperatures were generated due to adiabatic heating, in the same way that the air in a bicycle pump gets warmer as it is compressed by the pumping action. Put simply, as you push particles closer together, they heat up.

## Sinking metal

Gravitational loading was not the only way these larger bodies were able to heat themselves up. Most people are aware that the Earth hosts a metallic core at its centre, which is solid in the middle and surrounded by a liquid outer part. Yet, as we've seen, Earth is not the only planetary body to host a core. All planets have one, but they are not always the same composition as Earth's core and they are certainly not the same size either, with the largest, Jupiter's core, being tens of times larger than the total mass of the Earth. Even some of the smaller objects such as asteroids were able to form a core. The main requirement for core formation is that the planetary body had to be large enough to remain molten, or partially molten, for a reasonable time – at least for a few tens of millions of years. The process of forming a core itself has an important role to play in the overall heating history of a planet.

Without Earth's core – and, in particular, our liquid outer core – we wouldn't have a magnetic field. This acts as a protective shield around the planet, playing a vital role in diverting the solar wind. Without it, charged particles from the Sun would strip away our ozone layer, allowing harmful ultraviolet radiation to affect our planet, something that would not go down well with biological cells. In this way, we probably need to thank our core for our very existence on the planet, allowing for a biologically safe space for life to continue.

For a core of metal like Earth's to form, however, scientists calculate that the planetary body must have reached a very high temperature, high enough to melt iron. This is reached at a little over 1,500°C (2,732°F), at which stage the now molten iron separates from the melted liquid rock making up the rest of the planetary body. The iron, being denser than the mostly silicate rock, naturally migrates towards the centre of the body because of the influence of gravity pulling it down.

Despite the high temperatures required for core formation to occur, as we've seen, it is a normal part of the evolution of large planetary bodies. Venus has a core that is either still partially molten, or was in the relatively recent geological past, and there is evidence that Mars had one in the past, but that it has now probably cooled to solid metal. The NASA InSight mission was sent to Mars to look at this in more detail. If a planet is warm enough at the present day to have some liquid portion of its core, then it will host a magnetic field. So even if the other planets are no longer warm enough to have liquid metal at their centres, they may well have produced a magnetic field in their past. Of course, it doesn't mean that these places were, or are, capable of hosting life, as there are other factors at play that may have prevented life being able to take hold. Nevertheless, the protective benefit afforded by a liquid core in producing a magnetic field is certainly handy for life.

But we haven't yet addressed how the existence of a core, solid or liquid, has a heating effect on the terrestrial planetary body in which it is found. The heating effect comes about in two ways as a by-product of the core's formation. Firstly, the physical sinking of dense iron and nickel to the centre of the body generates frictional heating as the metal rubs against the stationary materials it sinks through. At the same time, the sinking process itself converts potential (gravitational) energy into thermal energy. This might not sound as if it could produce much of an influence on the overall temperature of a body, but scientists calculate that the temperatures generated could have been of the order of 1,500°C (2,732°F). This extra burst of heat helped to keep the outer surface of the planets molten for longer. This means that planets that were hot enough to form a core were then heated up once again by the process of forming it.

Nevertheless, Earth's primordial heat can only account for around half of the heat flowing from its insides to the surface at the present day. The mantle really has acted as a great insulator over the course of 4.5 billion years, storing that ancient thermal energy very effectively. But without our planet's other major source of heat, which we will discuss next, Earth wouldn't be the life-giving place it is today. This secondary heat source has also had a significant effect on all other terrestrial planetary bodies, and many beyond, playing a major role in their evolution through history.

**Nuclear heat**

When the rocky parts of planetary bodies were shaping up, they sampled the elements that were present in the protoplanetary disk from which they formed. This disk was the birthplace of everything we see in the Solar System today and contained all the ingredients the planetary bodies required, including the rock, metal and even organic building blocks for life itself. Once the planetary objects had eventually gathered the materials they needed from this disk, they found they had also captured some rather unstable, radioactive elements.

While this might sound rather ominous – maybe leading you to think of 'end-of-the-world' and nuclear fallout scenarios – within the terrestrial planets, these unstable elements were, and are, incredibly useful. Within Earth, they provide us with the other 50 per cent of heat our planet requires to be the perfectly cosy place it is at the present day.

The important thing about the radioactive elements is that their atomic nuclei are unstable because they don't have enough binding energy to hold the nucleus together, due to an excess of protons or neutrons. As they lose energy in attempting to become more stable, they release small particles – as alpha, beta and/or gamma radiation – that collide with other materials in the Earth. It is the energy of these particles travelling through the Earth that is converted to heat. Within Earth, the warmth provided, at present, from radioactivity mostly comes from the decay of just three elements: uranium, thorium and potassium. Yet these are stored in varying abundances within the mantles and crusts of all the terrestrial planetary bodies. If we knew the exact amounts of these elements within the planetary bodies, it would be possible to calculate their overall thermal budgets: how much heat and potential future heat they have. In reality, because they are large and complex structures with internal layering, it is not so simple to make this calculation because there is no easy way to determine the average composition of a planetary-sized object.

What we do know about our own planet – because we have had the advantage of being able to sample it in detail – is that the portions of crust that make up the continents are packed full of radioactive elements, making for a very concentrated region of heat production. The mantle, on the other hand, while having fewer heat-producing elements per kilogram of rock, makes up a much larger volume of the planet, so actually has a greater heating contribution overall.

Just like the heating effect provided by the sinking of iron through a planet to form a core, it might not sound as if radioactive heating could have played a significant role in a planet's overall heat budget. After all, it's just tiny particles transferring their minuscule amounts of energy into other parts of the Earth's interior. Yet, because the heat-producing elements are stored in such large abundance, they have the potential to release a great deal of heat between them. Perhaps more important, however, is the extreme length of time they've been actively decaying, which, in many cases, is of the order of millions or billions of years.

Let's take the example of uranium-238, which is the most common isotope of uranium found in nature. The '238' indicates the mass of the isotope, which is simply a particular variety of that element, with other isotopes having a different mass owing to a different number of neutrons within their nucleus. A range of uranium isotopes undergo radioactive decay at different rates, but if we focus on uranium-238, then it is known to have a half-life of around 4.5 billion years. This means that it takes, on average, 4.5 billion years for half the atoms in a collection of this isotope to have decayed. In doing so, they transform into thorium-234, a daughter isotope, which is also unstable, so it decays to another isotope, which is also unstable. The continuing process of the production of an unstable atom to another unstable atom forms a chain of decay, providing heating at every step.

Nevertheless, according to quantum theory, it is impossible to know exactly when an individual atom will decay, even if the atom has existed for a very long time. All we can do is to make statistical predictions based on a large dataset. It doesn't mean that exactly half the atoms of uranium-238 will have decayed after 4.5 billion years, because there is some spontaneity to the process, but it gives a very good approximation. Because it is such a long period of time, uranium-238 has been capable of delivering heat to our planet throughout the history of the Solar System. In fact, even by the time our Sun bids farewell (in approximately five billion years from now), only around three-quarters of the uranium-238 the planet started out with will have decayed.

But this is not the case for all the radioactive elements. Aluminium-26, for example, with a half-life of just 700,000 years, was almost all decayed away within about the first 100 million years after the formation of the planets. However, this doesn't mean that such shorter-lived isotopes have had a lesser effect on planetary evolution. During the time that aluminium-26 was active and generating heat, it was an incredibly effective source of planetary-scale warming. The heat produced was enough, in fact, to melt the planetesimals within which it was contained, turning their rocky interiors into molten rock, or magma. This was probably the case even on much smaller objects like the asteroids, such is the effectiveness of aluminium-26's heat production. Even if a small asteroid, perhaps just a few kilometres in diameter, had collected up a tiny amount of aluminium, say 0.005 per cent of its bulk, then this could easily have provided enough heating to melt the entire object, despite the fact that the vast majority of the aluminium is aluminium-27, which is not radioactive.

One of the conundrums to figuring out how much heat aluminium-26 produced in any given planetary body is that the isotope has all now decayed away, so we can't measure its abundance at the present day. Luckily, however, as it decayed by loss of a beta particle, it was transformed into a new isotope, magnesium-26, its daughter isotope. This one is stable and, as such, has been preserved in the planetary bodies it was formed within ever since. The measurement of this isotope within planetary bodies allows scientists to calculate exactly how much heat was produced in forming it. It is almost certainly the warming effects from the decay of aluminium-26 that provided enough heat for these bodies to melt iron and subsequently form their cores. As we've seen, without Earth's core and its magnetic field protection, we probably wouldn't be here. So, if we want to go back further, we need to thank aluminium for our existence.

Looking at the planetary bodies surrounding us, based on what we know about their composition, which varies from body to body, we can piece together at least some of their warming histories. The Moon, like all the rocky bodies surrounding Earth, contains radioactive heat-producing elements. We know this because we have been there and sampled lunar rocks that have been analysed back on Earth. The Moon is found to house considerably fewer heat-producing elements compared to our own planet, which might go some of the way to explaining why it is now dormant, being much colder than our own planet. But there are other factors at play, and we will focus on how these bodies cool in the following chapter. Mars is thought to be cooler at the present day for much the same reason. While results from the Viking landing sites suggest that the red planet contains heat-producing elements, it is depleted in them relative to Earth.

Scientists planned for the NASA InSight mission to focus on this theory as well. The idea was that the spacecraft lander would determine, by indirect methods, the heat flow of the planet. Instead of measuring the rocks themselves for their elemental abundances, it was designed to drill into the Martian crust, sinking a mole equipped with a thermometer. At the time of writing, it was determined that the mole was unable to reach its required depth for accurate measurements to take place. However, it is hoped that future missions will attempt to make the same measurements, allowing scientists to back-calculate the number of heat-producing elements Mars contains.

Looking towards the centre of the Solar System, we come across Mercury. This small planet, while being the closest planet to the Sun, isn't very hot. Its surface temperature can indeed rise to 400°C (752°F), but it can also drop down to -170°C (-274°F), and it has water ice at its permanently shadowed poles. One of the reasons is that Mercury doesn't have an atmosphere to trap heat, but a second reason is that it is thought to be dead inside, or at least without enough internal heat energy to power activity at its surface. Mercury has a big heart, with its core taking up around 85 per cent of the planet's radius, but the corollary is that it does, therefore, have a small mantle. Thus, it lacks a suitable storage environment for the radioactive elements that could act to heat it long term. Those silicate rocks really do act as a lovely warm blanket, so with its thin mantle layer, Mercury has failed to insulate its primordial heat as well as Earth, and has cooled much more quickly.

Mercury hosts a heavily cratered surface, which was created during the late heavy bombardment – a period of Solar System history around four billion years ago marked by many large asteroids and comets colliding with the terrestrial planets. On Earth, our more recent tectonic and volcanic activity acted to cover up the evidence of this period of Solar System frenzy, but the history is preserved much better on less active planetary bodies. Nevertheless, the measurements made at Mercury when NASA's MESSENGER spacecraft orbited between 2011 and 2015 found evidence that this little, seemingly dead, planet might still be warm inside, enough even to host a liquid core. The strange thing about Mercury's core is that it is thought to be solid on the outside before turning into liquid, then solid again at its very centre. Scientists were relieved to find that Mercury contained some liquid in its core because, like Earth, it has a magnetic field, albeit a weak one, and they were not able to figure out how it could produce one without the movement of liquid metal within its interior.

## Heating the ice

When we move out to the gas and ice giants of the outer Solar System, things are a little different from their smaller, terrestrial (rocky) counterparts. Firstly, it is hard for scientists to come to firm conclusions about the current state of these planetary bodies and, in turn, their history. This is directly related to their distance from us and the difficulty of reaching them to study them in more detail. In addition, they don't have solid surfaces, so they present extremely challenging environments for spacecraft, let alone humans. Nevertheless, several key space missions have studied these objects such that scientists have been able to make some important inferences about their current states, informing us about their evolution since formation. In particular, understanding whether they generate their own heat is and has always been an important scientific focus.

Let's take Jupiter as an example. As we travel further from the Sun, it is the first of the giant planets we meet, and it is also the largest object orbiting the Sun. In common with all the outer planets, Jupiter doesn't host a solid surface. In addition, its 'insides' are composed predominantly of helium and hydrogen. Nevertheless, Jupiter's 'surface' – defined as the point at which it attains Earth-like sea-level pressures of one atmosphere – sits at around -145°C (-229°F).

Its surface is cold simply because it is far from the Sun but, conversely, its interior is far from cold. Instead, the regions below Jupiter's surface are hot enough to turn its gases into liquid, and even plasma, as the core of the planet is approached. Of course, like all large bodies, pressure increases with depth and at nearly 140,000 kilometres (87,000 miles) in diameter, compared with Earth's relatively small 12,742 kilometres (7,918 miles), Jupiter has some extreme depths, helping to turn its gases into denser forms of themselves.

Normally we would class hydrogen as a non-metal but under the pressure hydrogen finds itself on Jupiter, it becomes so compressed that many more atoms can occupy the space normally filled by only a single atom. In this state, the atoms' nucleii are so close that they start to 'share' some of their electrons, which move from atom to atom and thus become a 'metal'. On Jupiter, this behaviour of hydrogen generates the planet's giant magnetic field, acting the same way as the metallic, liquid outer core of Earth.

But maybe Jupiter has more than just hydrogen and helium. At its very centre there could easily exist a core of rock or ice, or both, scientists have speculated. A core was always presumed to be there because – based on our accepted knowledge of planetary formation – without a large mass at its centre capable of attracting matter, Jupiter would have struggled to capture the gases it now hosts, which have enabled it to grow into the giant it is today. However, it could not be known for sure until 1997, when gravitational measurements were made of the planet, that Jupiter possesses a core. The size of this (possibly) rocky core is thought to be equivalent to 10 to 15 Earth masses. Jupiter's core is thought to sit at around 24,000°C (43,232°F), a temperature that could actually mean the core is, in fact, no longer present as the structured and discrete unit we think of when we imagine a planetary core. Instead, the extreme temperature could mean it has been melted to liquid that was subsequently mixed into the rest of the planet. In 2016, NASA's Juno mission revealed that Jupiter's core was indeed enormous but that it was 'fuzzy', supporting this 'melting' hypothesis and suggesting that its core might be partially dissolved.

With this large core, it might come as no surprise to learn that most of Jupiter's heat is thought to be primordial in origin: the warmth left over from its formation. But at the same time, it's generating some new heat too. This is from the movement of helium throughout the layer of liquid hydrogen surrounding its core – a layer that forms the majority of the planet's radius. The helium separates from the hydrogen and, being roughly four times its mass, tends to descend through it. Where it is cooler at shallower levels, the helium forms droplets that rain down through the hydrogen until they dissolve at deeper levels where it's hotter. As it sinks, the friction resulting from this movement between the two elements creates a source of heat. This might remind you of the heating effect that occurred on the terrestrial planets when their molten cores formed.

Moving on to Saturn, which, while sitting next to Jupiter and also hosting a potentially rocky or icy core that may be 10 to 20 Earth masses in size, we find that it has today lost much of its primordial heat. This may simply be because it started out cooler than the giant planet next door, so it didn't have as much heat to give, hence its potentially icy core. The heat that Saturn radiates today is roughly double the amount it receives from the Sun and is almost certainly generated in the same way as Jupiter's extra heat, from the sinking of helium through the liquid hydrogen surrounding the core.

Uranus is perhaps the odd one out of the outer planets. It radiates less energy than it receives from the Sun and therefore seems to be lacking an internal heat source. Its surface is approximately the same temperature as that on Neptune, which leaves scientists a little puzzled as it is also roughly the same size as Neptune, yet closer to the Sun. One explanation for Uranus' apparent lack of heat could be the planet's obscure orientation: it essentially orbits the Sun on its side (its poles are where we expect the equator to be). In addition, it has a chaotic and energetic atmosphere that may have dispersed its primordial heat early on. So, even if Uranus started out with significant heat from its formation, it seems to have lost it in the intervening years. Its core, composing 80 per cent of the mass of the planet but just 20 per cent of the radius, is probably made of water, ammonia and methane, compressed to the density of a liquid. A rocky core would be much better placed to provide heat.

Just like Jupiter and Saturn, Neptune – the furthest planet from the Sun at nearly 4.5 billion kilometres (2.8 billion miles) – radiates over twice as much energy as it receives from the Sun, meaning that it must have an internal heat source. Neptune's surface is roughly the same temperature as that of Uranus, despite it receiving just 40 per cent of the solar radiation, being a further 1.6 billion kilometres (1 billion miles) from the Sun than Uranus. Neptune's internal heat source is enough to drive the fastest planetary winds found in the Solar System and its current heat flow may be accounted for by the heat left over from its formation. In addition, movement of elements in its core, which sits at pressures twice that of Earth's core and at temperatures of over 7,000°C (12,632°F) could also create extra internal heat.

Now that we understand a little about how the giant planets do or do not generate heat, we can get on to the moons surrounding these fascinating bodies, as it is these objects that host volcanic activity. Some of these moons sound rather Earth-like in many ways, but these bodies, whether made of rock or ice, exist in very different environments from the planets of the inner Solar System. The result is that many have been able to host active surfaces for the entire 4.5 billion years of Solar System history. The way they have managed to produce their heat, as we have mentioned already, is by tidal heating, something that we can now delve into.

## Heating up next to a giant planet

As we have seen, the outer Solar System 'ice worlds' have rocky cores and it is expected that they were heated by these over the years. But the icy moons around them are generally small worlds and, as we'll learn in the following chapter, smaller bodies retain less heat than larger ones. They have less radioactive heat to begin with (because of a smaller core and subsequent smaller inventory of radioactive elements) and they just cool down quicker with a lower mass. So the expectation was that the small moons of the outer Solar System would all be frozen solid because they have run out of their initial heat energy. This is why it was such a surprise for scientists to find volcanic activity on ice worlds around Jupiter and Saturn, and ice volcanoes on places such as Pluto. Clearly, something was keeping the interiors of these worlds warm enough to fuel activity at their surfaces and, in many cases, maintain liquid reservoirs below their ice. This is where the idea of tidal heating came about, which we met in Chapter 6. Just a few weeks before Voyager encountered Io and discovered its volcanic plumes shooting into space, some scientists predicted that the little moon might be warm. They suggested that it was caught in a tug-of-war between the beastly Jupiter and the smaller but precisely timed pulls of two of its neighbouring Galilean moons, Europa and Ganymede, which orbit further away.

If you're a small moon sitting next to an enormous Solar System giant, then you can expect to be in for a rough ride as you go about your orbit. The scientists were, of course, probably correct, as there doesn't seem a better way to account for Io's high levels of volcanic activity. The gravitational tugging forces Io's rocky surface to bulge up and down by as much as 100 metres (328 feet) in a process similar to how Earth's oceans react to the gravitational effects of the Moon's orbit. However, on Earth the difference in height between a high and low tide is a maximum of around 18 metres (60 feet), and this is in water, which is obviously an easy material to move around. The forces required to deform Io's rocky structure by over five times this amount are clearly very impressive. The surface tides are made even more spectacular by Io's irregularly elliptical orbit, which acts to heighten the changes. As a result, Io's interior grinds against itself, which acts to kindle its fire. The tidal flexing and the resulting friction generate impressive amounts of heat, enough to melt rock and provide a near-perpetual magma supply. This makes Io one of the hottest and most volcanically active places in the Solar System. The thing about Io is that it is not a 'weird' outer Solar System object. It is actually much more similar in its overall composition and style of volcanism to the inner terrestrial planets and is even a similar size to our own Moon. Its crust is made of basalt and sulphur and it has a metallic core and mantle. This means that we can apply a lot of our understanding about the closer terrestrial planets, which are a bit easier to study, when we are trying to find out what is going on with Io.

However, scientists are still unsure as to which portions of Io's interior produce its impressive heat. Is it just the upper layers, its asthenosphere, or is heat generated very deep near its core? To try to answer this question scientists need to look at how the volcanoes on Io are distributed, how often they erupt and whether their eruptions are affected by nearby volcanoes. This will give them an idea about the magmatic plumbing systems for the volcanoes. Are they deep or shallow, shared between volcanoes or individual?

To understand what is going on we need to take a multi-pronged approach, combining Earth-based telescope data with images and data sent back from spacecraft, because there isn't one single technique that allows us to view the entire body. For example, Earth-based telescopes can't see Io's polar regions, but NASA's Juno spacecraft studying Jupiter gets to glimpse these parts of Io instead. Different studies have been completed that combine the datasets available to determine how many volcanic hotspots Io hosts, and where exactly they are found. There are some inconsistencies, but these seem to be related to the timeframe of when the observations were made, suggesting that Io's volcanic activity varies over time, which is not surprising for such an active body. Nonetheless, it seems that Io hosts well over 80 individual hotspots at any one time, and the studies conclude that its total heat output appears to be steady. These are important pieces of information, helping scientists to suggest that the most likely explanation is the heat production is coming from a shallow asthenosphere layer, which is thought to be around

50 kilometres (30 miles) thick and is expected to be partially molten: a magma ocean just like Earth's Moon had in the past.

But while Io is the only Galilean moon to host a fiery surface, indeed the only planetary body not within the inner Solar System to erupt hot rock at its surface, it is not the only moon still generating heat at the present day. Many of the others, particularly the others around Jupiter and those around Saturn and Neptune, also undergo tidal heating. While they have rocky cores that get heated, they are otherwise made of volatiles, in liquid and solid form, meaning that their volcanoes are of the cryovolcano (icy) variety.

## The oddball

In 2015 the NASA New Horizons mission passed Pluto, capturing images of the fresh, young, crater-free surface of this icy dwarf. The New Horizons spacecraft found towering mountains of water ice 3,500 metres (11,500 feet) tall, sitting above methane and nitrogen ice that formed smooth plains. These features strongly support the idea that Pluto has geologically re-worked its surface in the recent past. In fact, it could even be active today, suggesting it finds a way to warm its ice such that it can flow.

At the time, findings from this ground-breaking mission were a complete surprise for scientists, who were left puzzled yet excited by the implications. For this frozen dwarf planet to host such a youthful surface requires that it must have some source of internal heat. The problem is, Pluto is very small – being just two-thirds the diameter of our Moon – and so would be expected to have lost any primordial heat long ago. In addition, Pluto is very far from the Sun, so it receives almost no solar heat. As a result, Pluto was expected to be dead and should have been this way for billions of years.

But it's not just Pluto that throws up questions, because it hosts five moons. One of them is Charon, which New Horizons revealed also shows a relatively smooth, fresh surface. Being even smaller than Pluto, at about half its size, around 1,200 kilometres (745 miles) across, it should also be completely cold and inactive.

So where could these objects have gained their heat? They do not undergo tidal heating, as far as we know, as they are located very far from any large Solar System body. There is another option, that perhaps Pluto experienced a large impact in its past, which would be a useful way to explain how its moons came into being: as debris left over from a collision. As we've seen previously, impacts of a large enough size are known to provide a source of heat, with the kinetic energy of the collision being converted into thermal energy. The question is whether this can explain how these bodies found the heat to fuel their more recent activity. Probably not, say scientists. The problem is that the Pluto moon-forming impact almost certainly happened too long ago, during the early frenetic phase of planetary evolution, and therefore any thermal heat that resulted should be long gone. While it's possible for a more recent impact to have occurred and for residual heat to remain, to confirm this, scientists require more evidence. Large impacts of this type have not been very frequent since around four billion years ago but, nevertheless, they can't be ruled out.

Trying to account for Pluto's behaviour is tricky. We understand so little about these faraway icy objects because we've visited so few of them, and so briefly, via spacecraft fly-bys that it is hard to draw firm conclusions. Either way, when we talk about 'heat' on such objects, we shouldn't get confused with warmth. Pluto is completely frigid. Its freezing surface temperatures mean that water exists in the form of ice and it is hard – rock solid – so doesn't move very easily. In fact, ice is Pluto's bedrock. The fact that its water-ice mountains tower above smooth plains suggests that below them there must exist a layer, or layers, of other more volatile ices: those that melt more easily. This is similar to the magma ocean asthenosphere layer that we discussed on Io, except it's not made of molten rock but instead composed of methane, nitrogen or carbon dioxide. If Pluto represents a dynamic system, as suggested by its smooth, young surface, then there must be a way to replenish these surface layers from below. Despite the fact that no activity such as cryovolcanoes or plumes have been spotted on Pluto yet, it doesn't mean they're not there – we just haven't looked enough – and their presence, if detected, would help scientists to piece together the state of the planet. So, the investigation continues.

However, there is another potential mechanism for moving stubborn ice around and it comes back to something we mentioned much earlier in relation to the fundamentals of how we make rock melt: by adding a substance that lowers its melting point. Like adding salt to ice on a driveway, or water and carbon dioxide to the peridotite mantle, on Pluto the water ice may be able to flow more readily by mixing with other compounds such as ammonia. As we'll see in Chapter 11, ammonia has also been detected on Pluto's surface and scientists suggest that just a 5 per cent mix of it into water ice could make the ice flow like a slurry, similar to a rocky lava flow on Earth but at freezing temperatures.

Within the last few decades, what we've discovered about the planetary bodies within our Solar System is that their evolution is not complete, because they still retain some warmth, in varying amounts, and many even find ways to generate new heat. They may have been around for nearly five billion years but the many ways they've found to heat themselves has allowed for a long and interesting geological history which, for some, is still going strong. But heating up is only half the story, because what really matters is how they lost their heat, which is even more significant to their evolution.

# CHAPTER EIGHT

# Cooling Off

W e've seen that many of the planetary bodies within the Solar System, particularly the terrestrial ones, didn't have any trouble finding ways to warm up early in their history. However, maintaining that heat for a sustained period of Solar System evolution was, in many cases, much harder. The way in which a planetary body cooled and, in particular, how quickly or slowly, has a substantial bearing on its evolution, including whether it was able to host liquid water, so giving it the potential for habitation. How a planet cooled down after its frenzied birth is unique to the individual object, because it is based on a whole range of factors, the major ones being:

- how and where the object formed (that is, how far from the Sun),
- what it contained from the start in terms of the particular concoction of elements it collected up from the protoplanetary disk,
- what kind of impacts it underwent after formation,
- and its overall size.

Planetary bodies cool by conduction and/or convection. Convection can only take place if they contain liquid on their insides, or if their solid internal materials can be said to flow as if they are liquid. This is obviously the case for Earth's mantle, which sluggishly moves the planet's internal heat around and, in turn, moves the tectonic plates at the surface. Conduction, however, can occur even if a planet has cooled past the point of being able to move its insides, and this can take place even where there are no plate tectonics. While Earth may cool primarily by convection, those that cool by conduction, such as the Moon and Mars, are described as 'stagnant lid' objects, with no plate tectonics.

When it comes to a rocky mantle, whether it is warm enough to flow or is cooler and solid, it is made of silicate rock that is, as we saw earlier, a really poor conductor, so it's good at insulating the heat pumping out from the core. Thus, a large planet with a lot of rocky mantle tends to cool slower than a smaller planet, or one with a small proportion of mantle.

The problem with a planetary body cooling too quickly is that once its insides have passed the point where they are no longer able to convect, the planet won't be capable of fuelling significant geological processes at its surface. A planet that is cold inside, with no volcanism, is a dead one. Nevertheless, without planetary-scale cooling in the first place, we would never have seen volcanic features on these bodies either. In terms of making a life-giving, geologically interesting planet, the planet must cool at just the right rate such that a sustained level of volcanic activity can be maintained, and not too quickly that the activity dwindles away and the planet dies.

We saw in the previous chapter the case of Mercury, which is no longer warm and active. Scientists are still trying to figure out exactly what happened to Mercury in its history, but MESSENGER shed some light on this small and dense mysterious planet. The reason scientists think Mercury ended up with so much core and so little mantle is that it probably experienced a large impact early on in its history. This random blow from a space object is purported to have ripped off its outer mantle shell, which was expected to have been much thicker, forming a larger proportion of the planet at the time, and left behind just a little of the mantle along with the sturdy core. In terms of location, it seems that Mercury might just have been in the right place at the right time (or the wrong place at the right time), and was unlucky to have experienced such a large collision early in its evolution: a collision that for ever affected its future. But Mercury was not alone in experiencing very large impacts and its mantle-ripping collision was not even as dramatic as Earth's big smash, which is thought to have vaporised our planet and forged our Moon. If the object that collided with Earth had hit at a slightly different angle or speed, then it could have ripped off our mantle in the same way as Mercury's. Luckily for Earth, while it went through a phase of great change, it managed to gravitationally hold on to most of its mass after briefly being ripped apart, and it gathered itself back together again, leaving just a little bit of material for its now-important Moon.

## Size and location

The total amount of heat a planetary body can hold is dependent on its volume, but the rate at which that object cools is determined by its ratio of surface area to volume. The surface area to volume ratio is smaller for larger planets so, all else being equal, a larger planet cools more slowly than a smaller one. If we think about the Earth, Mars and the Moon then, ignoring any internal heat production from radioactive decay, we can conclude that the Moon will cool twice as fast as Mars, which will cool twice as quickly as the Earth. This is simply related to their relative sizes. Of course, other factors come into play. The Moon also has fewer heat-producing elements than the Earth, being just 2 per cent of the mass of the Earth, so it was unable to replace its lost heat as efficiently as our own planet. Subsequently, the Moon is long-dead whereas Earth is still flourishing.

Asteroids, being much smaller than their full-size planetary counterparts, lost their heat relatively rapidly too. But in many ways, this was actually quite a useful outcome. Their rapid cooling directly affected their subsequent evolution. In cooling and becoming a solid body, they were halted in their evolution very soon after they formed, on geological timescales anyway (millions of years). This means that asteroids can play a vital scientific role at the present day, acting as time capsules from the period of planet formation, preserving a record of the Solar System conditions at the time they formed, cooled and solidified. Conversely, the larger terrestrial planetary bodies that stayed warm for longer periods continued to evolve physically and chemically during this time. For many, this means that they even experience changes up to the present day. This makes it much harder for scientists to unravel the planetary histories and delve back in geological time to when they formed. Without being able to piece together how these objects started out, it is very hard to answer questions about how and why they became the planets they are today, particularly if we want to understand what it takes to make a planet habitable. Thankfully, by studying the asteroids, we can peer into this period of planetary formation to understand an epoch of Solar System evolution that is no longer preserved.

With just a slight difference in cooling rate playing such a determining role in how a planet evolves, it's no wonder that the planetary bodies surrounding us are all so different. Despite Mars being one of Earth's neighbours, inhabiting a similar region of the Solar System, its faster cooling rate compared to Earth's played a pivotal role in its evolution.

This extends even to the present day, with repercussions on whether it could sustain life. Mars' smaller mass compared to Earth's – about ten times less – had a two-fold impact on its cooling and evolution: it doesn't possess as much gravitational energy, and it is also expected to host fewer radioactive elements. The latter meant that Mars was unable to generate as much heat as Earth, so it cooled more quickly and, consequently, its molten core solidified.

Mars' magnetic field was, therefore, short-lived in comparison to our own planet, with scientists suggesting it was gone 4.2 billion years ago. A knock-on effect of this was that Mars' atmosphere was not protected from the solar wind and so its atmosphere was stripped away. With little or no atmosphere, there is no hope for retaining liquid water on the surface of a planet. This meant Mars was unable to hold on to the surface water that it had built up. There is lots of evidence suggesting that the surface of the red planet was once covered with running water.

The relatively rapid cooling of Mars also had another effect. It developed a thick, rigid lithosphere which meant that plate tectonics was impossible. Even if plate movements had developed at some point in its past, there was not enough heat to sustain flow within its mantle and the lack of water meant that the lubrication for the process was also missing.

All of these factors brought about by the relatively fast cooling of Mars were the reason it was probably unable to support life. If it did ever host organisms, then it is hard to understand how they would have survived without a protective magnetic field unless they were underground, and, importantly, how they would have managed without a molten interior to support plate tectonics. Plate tectonics drives the long-term processes that support the surface activities to encourage life. Nonetheless, despite the lack of plate tectonics, Mars' surface was still able to host active volcanoes, since the insides of the planet were warm enough to convect for a reasonable period of time. In fact, Arsia Mons – a shield volcano we met earlier, just south-east of Olympus Mons – may have been active as recently as 10 million years ago. Despite no current evidence to support the concept of Mars hosting life, it is possible that this early 'non-plate tectonic' activity could have allowed life to develop in the past. It is rather speculative perhaps, but scientists have shown that the red planet hosted the right conditions for life in the past – it certainly had liquid water at its surface – so future missions will be searching for evidence of life having evolved there.

Even looking at this one example goes part of the way to demonstrating how delicate the balance is in the formation of a planet just like Earth, one that hosts the right conditions to create and sustain a habitable surface environment over billions of years. In addition to a planet dissipating its internal heat at just the right speed to provide a suitably active surface, the other key control on habitability is location. Habitable planets like Earth are often said to be in the 'Goldilocks zone', not too close to and not too far from the Sun, sitting in the Solar System's sweet spot for potential liquid water and life. So, even if a planet meets all the other desirable conditions for hosting life, such as having been created from the right starting ingredients and retaining enough heat for a reasonably long period of time, it is thought that it still needs to be located in the right zone of the Solar System for that life to thrive.

Venus is also generally considered to be in the Goldilocks zone, but it doesn't demonstrably host life on its surface at the present day, and there is no evidence it ever has done, thanks to its toxic atmosphere and very hot surface. As we've seen, despite their differences, Earth and Venus are often referred to as 'twin-like' planets because of their similar size and distance from the Sun. As a result, we'd expect them to have cooled at similar rates. Yet, rather confusingly, they have cooled quite differently. Venus, despite being closer to the Sun and having a hotter surface, cooled down much more quickly. In fact, Earth's 'evil twin' is thought to have passed through the minimum temperature required for convection to take place within its interior around two billion years ago, whereas Earth is still to reach that point in its cooling history. Scientists suggest that this could be related to the different mechanisms by which the two planets cooled down: Earth being controlled by plate tectonics, and Venus, being a 'stagnant lid' planet just like Mars, lacking plate tectonics. But the question is then: why did they cool by different mechanisms when they started out so alike?

The reason might be related to the absence of water on Venus, as this is thought to be a key factor controlling whether plate tectonics can take place. Basically, water is thought to act like a lubricant allowing plate tectonics to occur. It is the oil in the Earth engine. Water lowers the viscosity of rocks such that they are easier to move, while also reducing the friction of brittle rocks, helping them to slide past each other more easily.

When Venus formed, it should have collected up roughly the same starting ingredients from the protoplanetary disk as its Earth twin, so it should have had the same amount of water. Even if that water was lost early in its history, as could have also happened on Earth, the two planets are expected to have experienced roughly the same number of impacts from space, which is the other way they could have received water from comets and asteroids later on. So, what happened to all of Venus' water? This is where a planetary body's distance from the Sun may play a key role.

The surface of Venus, thanks to it being a third closer to the Sun than the Earth, remained just that little bit warmer throughout its early years. It is thought to have retained a magma ocean surface for longer than Earth has: hundreds of millions of years as opposed to tens of millions of years in the case of Earth. Its slightly slower rate of cooling meant that any water present in its atmosphere, having been released from its interior by volcanic degassing, was unable to condense out into oceans, as occurred on Earth with its slightly cooler and more stable solid surface. As a result, Venus' atmospheric hydrogen, being packed with more energy because of the heat, escaped into space, leaving behind a dry planet. While this is only one hypothesis, and there may well be other explanations (for example, Venus also rotates a lot slower than Earth, which may also have affected its cooling rate), it seems a plausible mechanism to explain Venus' hot and dry atmosphere today. Its current atmosphere is composed of around 0.002 per cent water vapour, compared to Earth's 0.4 per cent and, of course, Earth – thanks to its cooler surface – hosts a lot more water on the surface, a factor that has almost certainly helped start and sustain life.

Understanding the conditions required for a planet to host life is complex. When scientists look at planets in other star systems far removed from our own, they can use the knowledge gained from our Solar System to help focus their search for habitable planets. But the issue is that we only have our Solar System for data right now, and Earth provides the only data point for confirmed habitable planets. Recent studies suggest it is possible that we might need to extend the Goldilocks zone to both closer to and further from the Sun, as the evolution of a planetary body is dependent on so many more factors than its distance from the Sun. Just because we haven't found life elsewhere yet doesn't mean it's not there.

As we've seen, there are more volcanically active moons in our Solar System than planets. This might seem counter-intuitive based on what we've learnt previously, as moons, being relatively small, supposedly cool faster. However, while primordial heat and the heat produced by the radioactive decay of elements within planet or moon interiors are important ways for some of these bodies to remain warm, for the moons surrounding the largest planets in the Solar System, factors like tidal heating can play a very important role. This is where a planetary body's location in the Solar System can change everything. Now, when searching for the potential for life elsewhere in the Solar System, some scientists are more excited by several of the moons surrounding giant planets, as opposed to Venus and the seemingly dead Mars in the so-called 'Goldilocks zone'.

## Making solid ground

The outer surface of a planetary body reveals its most recent phase of evolution and geological activity. For planets that are no longer active, their 'most recent' phase of activity could be billions of years ago. Yet, even on a continuously active planet like Earth, one which has never stopped re-surfacing itself thanks to the processes of plate tectonics, some portions of crust can still be billions of years old. These pieces of ancient material just happen to have escaped and survived the never-ending reprocessing of the planet's surface. For this reason, Earth's crust documents bits and pieces of a long and complex history going back at least four billion years. It has preserved crust from various stages of its evolution. Our outer shell, therefore, reveals a jigsaw picture of past events, but all the pieces have been jumbled up. If we can pick them apart and piece them back together in the right places, then we can start to understand what is going on, and what occurred previously beneath the surface. Of course, some of the jigsaw pieces are missing, having been destroyed by plate tectonics; Earth simply hasn't preserved a record of all the events in its ancient history.

But how do we go about making a crust in the first place? Well, obviously we need to cool a planet down, as a molten ball of rock can't produce a solid surface. The extreme heat generated within the terrestrial planetary bodies early in their history meant, understandably, that they spent a reasonable amount of their early lives as molten bodies, or at the very least, partially molten bodies. Here we're talking, in many cases, from tens to hundreds of millions of years. These planetary bodies were exceptionally warm from the start, yet they were sitting in the cold vacuum of space, and their heat only wanted to move in one direction, towards the cold. Eventually, this cooling led to them forming a hard, outer shell, a crust of solidified rock.

For a planet to start forming a solid crust, it has to cool to the point that allows its molten rock to crystallise and solidify at the surface. We could just give a planet a bit of time and eventually it will cool off enough to form a crust, like a skin on gravy, but there were a few processes during planetary evolution that sped up the cooling of the planets at key points. In nearly all cases, the first crust a planetary body makes is preserved only for a short time because it is either destroyed by a huge impact or covered over with new crust, thereby being re-worked and melted back into the interior of the planet.

As we've learnt, one of the consequences of having a hot, molten rocky planet is that it tends to gravitationally segregate into internal layers, with the metallic core sinking to the middle, surrounded by a rocky, molten mantle, the magma ocean. While we saw that the formation of a core generates heat within a body, we didn't address how it also brings about cooling because it displaces portions of mantle through which it descends. This shifting of a planetary body's insides is thought to have generated widespread and rapid convection within its magma ocean and, in turn, brought about the first major stage of cooling and degassing from its interior. From this process, many of the planetary bodies created their first transient crusts. But these 'gravy-skin' crusts didn't usually last very long at all and were efficiently melted and mixed back into the boiling magma oceans, possibly as a result of further impacts from space.

Nevertheless, the planetary bodies continued to lose heat and were able to start making a better, more stable version of crust through the cooling of their magma oceans. These 'proper' primary crusts were not simply a solid film at the surface of a pool of molten rock; they were instead made of solid minerals that separated from the magma ocean as it cooled, and floated their way to the surface. The process of forming the crust is the opposite of forming a metallic core, where the dense minerals separate out and sink. Instead, the minerals that formed the early crust were less dense, compared to the other minerals and silicate melt that made up the magma ocean. As such, these lighter minerals migrated upwards through the relatively dense liquid–crystal-mush. This is all part of the continued phase of planetary differentiation where an object separates into layers of different composition, determined by how they are affected by gravity because of their relative densities.

Such a crust formed as the molten Moon, which had collected up and condensed material from the vaporised detritus of the Earth after it was smashed to pieces during a large impact, cooled and underwent its own crust formation. The great thing is that you can see the Moon's primary crust from Earth. If you've ever stared at the lunar surface in the night sky, then you will almost certainly have noticed it is made up of some patches of light materials and some patches of dark materials. It is those lighter areas that are the lunar primary crust, which may extend down to 50 kilometres (30 miles) in depth. These are the famous lunar highlands or 'terrae', which Apollo 16 explored in detail. The lunar highlands form around 84 per cent of the lunar surface. They are composed almost entirely of a single mineral, feldspar, which makes up a rock known as anorthosite. This near-white mineral is a silicate containing large amounts of potassium and sodium. The high abundance of feldspar in the anorthosite is what gives the highlands their light colour.

The rocks that form the lunar highlands are incredibly long-lived, surviving on the Moon for over four billion years. Scientists have no reason to think that primary crusts on other terrestrial planets wouldn't have formed in the same way. It's just that investigations haven't confirmed whether such crusts are still in existence at the present day. After all, the rocky planets were all formed originally from the same Solar System ingredients and were all molten bodies for millions of years, enough time to allow crystals to form and float to the top of their magma oceans. The difference is that they cooled at different rates and experienced varied histories after forming their first solid surfaces. The Moon, being relatively small, cooled quickly enough that it didn't experience a very long geological history of activity compared to the planets that surround it, and thus, its primary crust is preserved to this day.

The fact that the Moon has preserved its primary crust for such a long time is fortunate as it is the only planetary body we've sent humans to explore and sample. In doing so we've been able to learn about the early phase of planetary evolution throughout the entire Solar System, not just on the Moon. Without examining the Moon, we would struggle to understand what primary crusts looked like on the other planetary bodies where they no longer exist, or even how they formed. There is huge value to be gained from looking out into the Solar System to learn about our own planet. It is like some of our missing jigsaw pieces can be found sitting on the surfaces of other planetary bodies.

It's possible, yet still hotly debated, that Mercury and Mars might also retain some of their primary crusts. Further missions to these planets, including the return of rock samples to Earth, will be required to figure this out for sure. But these are larger planetary bodies and so they have experienced a longer history of activity than the Moon, because they cooled more slowly. It is their subsequent activity that can destroy the older materials. However, the reason it is likely that these planets have preserved portions of their primary crusts is because, as far as we know, they haven't experienced millennia of plate tectonics. Plate tectonics keeps Earth's geology interesting, but it also has the effect of erasing part of our geological history.

## Planetary face-lift

What about the dark patches on the Moon that appear in stark contrast to the lighter regions making up the lunar highlands? When early astronomers first studied the Moon in detail, these regions caused some unwarranted excitement, leading to their current name: mare. The dark patches represent the Moon's other major type of crust: the basaltic maria. Mare (pronounced 'ma-ray', the singular of maria) is

Latin for 'sea', which is exactly what the early astronomers thought they were seeing: huge bodies of water. With time, and as telescope design improved, we recognised that the maria were just darker regions of rock and, perhaps rather disappointingly, not oceans. Nevertheless, the maria are very similar to Earth's basaltic oceanic crust, except they formed as subaerial (above-surface) lava flows as opposed to underwater. These flood basalt eruptions potentially started as early as four billion years ago but continued up to as recently as one billion years ago, with the majority erupting around three to 3.5 billion years ago.

The fact that the basaltic flows had a relatively low viscosity meant that they were able to flow and enter depressions in the lunar surface, filling up more ancient craters that had resulted previously from impacts. These flows represent the Moon's secondary crust, and it happens to be a common crustal type on many other planetary bodies and one that is often preserved to the present day. The secondary crusts on the Moon, Mars, Earth and Venus were all formed after the crystallisation of primary crusts from their magma oceans. These usually basaltic crusts form from the partial melting of planetary interiors rather than just the solidification of them. On Earth, our secondary crust is produced by mid-ocean ridges and mantle plumes, but on other rocky planetary bodies it is mostly made by mantle plumes forming flood basalts, since plate tectonics is not a feature.

Until now, I've only focused on the crusts of the rocky bodies that experience especially high temperatures. Yet the icy moons of the outer planets also host secondary crusts, except that they are made of different forms of ice, not rock. As we'll see in Chapter 11, Europa's icy crust might even undergo some form of icy plate tectonics.

While primary and secondary crusts can account for most of the surfaces we can see on planetary bodies throughout the Solar System, Earth is unique in possessing another type of crust, known as tertiary crust. This is our continental crust, and it can only form because we experience plate tectonics; it is formed from the re-working and melting of the secondary basaltic crust when it was recycled back into the mantle at subduction zones.

Continental crust forms the thick portions of the outer shell of the planet that make up the high mountain chains, including some of the highest and most spectacular volcanoes of the steep-sided variety. But what is not obvious at first is that these mountains are a bit like icebergs, being grounded and stabilised by deep keels of rock foundation below, protruding down into the mantle. Wherever you see a large mountain, it has as much rock below the surface as above, balancing out its weight.

These portions of elevated continental crust are only able to grow so high because they are made of different material than that in the basaltic crust. Continental crust is less dense than basaltic oceanic crust, and both are less dense than the mantle from which they formed. As a result, both sit atop the mantle, but continental crust can grow the highest simply because it is the least dense of all.

The reason that the continental crust is different in composition from the basaltic magmas produced from the mantle is that, as we saw in Chapter 3, every time we melt the mantle, we change it. In melting the mantle via the process of partial melting – where the whole thing isn't melted – certain chemicals remain in the solid rock left behind in the mantle and others go into the melted part. The less dense chemicals enter the melt and progress to make a magma. If a solid rock is then produced from this, which in turn gets subducted back into the mantle and is re-melted to produce a new magma, an even less dense melt than the original one is produced, or produced in the first phase of partial melting. Basically, every time the Earth melts, the magma it produces gets more refined. The process of forming continental crust can be likened to a continuous distillation with the newly produced tertiary magmas being described as more differentiated: they are more silica-rich for one thing. Put simply, it turns basalt into granite.

This means that just observing the presence of a granitic crust at the surface of another planetary body would provide us with evidence for the existence of plate tectonics. The existence of tertiary crust can tell us a great deal about the history of that planetary body and allow us to take some educated guesses as to what is contained within: whether it is still hot, how quickly it cooled and whether its insides are molten or solid.

It is clear that the warming-up and cooling-off of a planetary body is controlled by a great many factors but it is important – if we want to understand the evolution of a body – that we get a handle on the history of these processes. How long a planet has taken to cool down and how this happened has important consequences for its volcanic activity, which in turn controls features related to its atmosphere and other surface processes, including the survival of water and even life. It's been said many times above, but it's worth stressing again, that an active planet – whether hot or cold – is a potentially life-harbouring one. But even an active planet can be biologically barren. If we want to understand how and why we live on a beautiful flourishing planet, we need to look in more detail at those around us to see what they did differently.

# CHAPTER NINE

# Fiery Moons

You might wonder how one person can safely stand next to an erupting volcano, watching lava flow past just a few metres away, while someone else's brain can be turned to glass 10 kilometres (6 miles) away from an eruption. What about the famous Icelandic volcano whose eruption in 2010 grounded aviation throughout much of the northern hemisphere? Eyjafjallajökull proved that we can feel the after-effects of an eruption very sorely even when we are hundreds of kilometres away. Despite the amount of observation and research carried out on Earth's volcanoes, we still struggle to predict when one will erupt, how long the eruption will continue, and how far-reaching and damaging it will be.

Yet most of Earth's volcanoes – those sitting above the sea anyway – erupt into roughly the same environment. These are conditions we understand very well in most cases. On the Earth's surface, we only experience very small differences in local atmospheric temperature and pressure. This means that, in the event that scientists have managed to understand the inner workings of a particular volcano and figured out exactly when it will erupt (which is not exactly an easy thing to do), they are able to predict, in theory, what the lava flows or ash clouds have the potential to do on eruption, including how high ash might travel, and how far and in which direction the lava and ash might spread.

But if we look at Earth's volcanoes that erupt under the sea, then we start to enter an environment that feels a little more alien, and one we don't understand as well because of its relative inaccessibility. Basalt erupting hundreds of metres below the surface of the sea behaves in a completely different way from the very same basalt erupting above the surface. The former rapidly succumbs to the extremely cold temperatures and high confining pressure under a deep column of water, meaning the lava can travel almost no distance at all, being halted just as soon as it is erupted. As we've seen, such lavas become rapidly quenched into a pillow-lobe. These inflated balloons of lava have exteriors that cool and solidify almost instantly on contact with the cold ocean, while their insides continue to grow with molten lava flowing in, insulated by the solid crust, until the lava bursts out of the lobe to create a new one. The same lava above the sea could travel for hundreds of metres before it cools and solidifies, having the ability to flow like a river in some cases, rather than ooze out like old toothpaste. Such a difference in eruption environment can result in deposits that look nothing like each other, despite both being produced by volcanoes on the same planet, made from magma of the same composition.

When moving out into space, things can get even more complicated. We have a good handle on the surface conditions of many of the planetary bodies within our Solar System, and from this we know that the environmental conditions can differ dramatically from those on Earth. The especially tricky part, however, is not knowing how a magma will react to the differences in temperature and pressure it will meet on eruption into these alien environments. While some of the effects are predictable – hotter temperatures make a lava flow further, for example – others are harder to predict.

Putting these differences aside, there are other factors that control how a volcano on another planetary object will erupt. These are mainly related to the magma itself. What is the magma made of and how did it form? Magma chemistry is controlled by planetary evolution, including what the planetary body contained when it was born 4.5 billion years ago, and how big it is. As we explored in the previous chapter, these factors control the speed at which a planetary body cools, which in turn affects the way its insides segregate and evolve over time. All of this controls how its rocks look and behave and what happens to them at the surface.

## Our weird planet

Arguably, Earth's most impressive volcanoes are the subduction-related ones. These are often steep-sided, high, sometimes ice-capped mountains that can cause all manner of destruction and chaos when they erupt. These often beautiful volcanoes have no analogues in space. The reason, as we've seen, is because subduction-related volcanoes require the subduction of one piece of lithospheric plate beneath another, which can't happen without the surface of a planet being brittle enough to be broken up into pieces that can move around in relation to each other, 'floating' on an asthenosphere below.

Thanks to plate tectonics, Earth's volcanoes often produce another very recognisable feature: linear chains. We have lines of volcanoes along spreading ridges (such as the Mid-Atlantic Ridge) and others forming chains overriding subduction zones (for example, the Aleutian Islands in Alaska). Even our hotspot volcanoes, which don't *require* plate tectonics, form linear chains as they erupt through lithosphere that is gradually moving (as in Hawaii).

In contrast, extraterrestrial volcanism rarely forms linear features and in the rare event that it does, they are not related to plate tectonics. But that doesn't mean we can't compare what we see on our planet to other planets, and it certainly doesn't make the alien worlds that surround us any less volcanically interesting. In fact, a lack of linear volcanic chains and steep-sided volcanoes reveals key information about the history of these worlds that can sometimes inform us about our own past.

While there are many volcanic differences between Earth and the other planetary bodies, the common feature is mantle plumes. In fact, looking at the inner Solar System, to put it bluntly, they are everywhere. These columns of hot rock rise from a planet's interior, promoting melting in the upper mantle to produce erupting lavas at the surface that are often voluminous, and sometimes absolutely massive. Luckily, mantle plumes are a geological feature that we have in abundance on Earth and so we can use our knowledge about our own mantle plume volcanoes to learn about those in space. But we must keep certain things in mind. Most obviously, as we've seen with Mars, its volcanoes were able to grow much larger than our own because the lava pumped out at the surface by mantle plumes just piled up in one spot for millions upon millions of years, without plate tectonics dragging the plate away to form a chain instead.

In this and the following chapter we will focus on the terrestrial silicate planetary bodies – that's the rocky ones that formed in the same neighbourhood as Earth, such as Mercury, Venus, Mars and the Moon. However, we'll also include Jupiter's moon Io, which, despite having formed much further afield, shares more similarities to these inner Solar System worlds than the icy and gassy ones that surround it. Despite their differences, this group of bodies have all produced volcanoes at their surface made of hot rock and, therefore, they are similar to Earth.

First, we will start with the smallest of these worlds, but otherwise two wildly contrasting celestial bodies: the Moon and Io. Despite being similar in size, the Moon is long dead while Io is the most volcanically active object in the Solar System. We will also look at the potential for a volcano on an asteroid, at the very 'small object' end of the Solar System scale.

## The dead one: the Moon

When exploring the Solar System's volcanoes, the most sensible place to start is the Moon, not only because it is our closest neighbour, but also because it literally formed from the Earth itself. The Moon is the planetary object we've explored in the most detail, having sent humans equipped with rovers to roam and scour its surface and collect hundreds of kilograms of samples to bring back to Earth. What scientists have managed to glean from the extensive studies they have completed, whether by telescope from afar or by analysing lunar rock samples, has told us a great deal about the geological history of the Moon. These scientists have generally agreed that the Moon formed during a giant impact of a Mars-sized object into our infant planet. They've discovered the ages of most of the large craters that pockmark its surface. They've been able to trace lunar history in detail, from the moment of the Moon's conception to the present day, understanding where much of its present-day surface came from. Yet, despite having all this information about the lunar surface and the geological history of the Moon in general, there is still a lot to learn. One of the reasons for this is because, even after all those visits we've made, a huge region of the lunar surface remains unexplored and unsampled. The far side of the Moon has only recently experienced its first soft landing, by the Chinese Chang'e-4 mission.

Often referred to with the misleading title the 'dark side', the far side of the Moon isn't always dark. Certainly, it is always turned away from us because the Moon is in a synchronous rotation with Earth, so we could perhaps call it 'dark' in the sense that it is unexplored, being one of the only places in the Solar System that we can't see directly from Earth. In reality, the far side receives as much light as any other part of the Moon over any given month. Nevertheless, for the more unexplored regions of the lunar surface scientists must piece together the history by extrapolating from those parts of the Moon that have been studied and understood in more detail. In the case of the Moon this is difficult because, from what scientists have been able to discover from spacecraft in orbit, the far side and the near side differ in some very important ways.

We think the lunar surface has been quiet and volcanically dead for a very long time. As we saw in Chapter 8, being smaller than Earth and not containing as many heat-producing radioactive elements, the Moon cooled faster and, as such, has a shorter volcanic history. But when we talk about lunar volcanics, we are mainly referring to the near side of the Moon. When the two hemispheres are compared, the far side appears very different. It is much more heavily cratered, indicating that its surface is ancient, having had more time to accumulate impacts from comets and asteroids. It is also thought that only around 1 per cent of the rocks on the far side are volcanic. It's not well understood why the two lunar hemispheres are so different, but it may be related to an uneven distribution of heat-producing elements within the Moon's interior. If there is a higher abundance of them on the near side, then over time they would have provided more heat in this region, keeping volcanoes flowing for longer and maintaining a younger surface. But other theories relate to a large impact from space occurring soon after the Moon formed that excavated the far side. While this seems to be inconsistent with geochemical studies, we still need to find an answer to this question.

Despite the Moon mainly hosting volcanoes on only one of its hemispheres, the near side, and volcanoes that don't instantly seem to resemble our own, it has still produced some interesting volcanic features. While sometimes leaving scientists rather perplexed, these features have enabled insight into the history of our own planet.

The lunar volcanic deposits cover just over 30 per cent of the nearside surface but, despite their limited size, they reveal information about a large portion of the geological history of the Moon in general. The basaltic maria are lava flows that, in many cases, appear to be closely associated with basins that were formed by large meteorite impacts. In fact, it was originally assumed that the lavas within the craters were produced directly by the impacts themselves, with the energetic force of the collision melting the crust and producing molten rock within the basin. However, it was later decided that the lavas flowed some time after the impacts had occurred, and that they simply filled the depressions. It could be that the basins merely facilitated the eruption of the mare owing to the removal of the overburdening rock during the impact, and by opening up fractures which the magma could move through to the surface. Either way, the maria are related to the basins, and the fact that they sit within them is very telling. It shows that the basaltic maria flows must have been very fluid, runny enough to spread quickly before cooling and solidifying and, therefore, flowing a great distance. One of the reasons scientists think this was possible is because lunar gravity is just one-sixth that of Earth's. This lower gravity results in taller lava flows compared to those on Earth, which cool slower because they have a reduced surface area to volume. They can, therefore, travel further.

But it is hard to put into context the scale of these flows without comparison to basaltic flows on Earth. Even the huge volumes of basalt that were erupted in one of the most recent phases of activity at the Kilauea volcano on the island of Hawai'i in 2018, which caused over $800 million of property damage, hardly compare. These extensive lavas were part of the lower Puna eruption, and they flowed rapidly in apparent lava rivers, engulfing hundreds of roads, homes and businesses. The event was relatively short-lived – it was all over within a few months – yet the lava covered 35 square kilometres of land (13.7 square miles) as well as creating 3.54 square kilometres (1.37 square miles) of new land in the ocean.

If we compare this to the Mare Imbrium, a single lunar mare that erupted over 2.5 billion years ago, then we can see the contrast in size. This flow extends over 1,200 kilometres (750 miles) from its inferred vent within the Imbrium Basin. This is comparable to the distance from London to the south of France! As such, the basin within which this flow is located can be seen with the naked eye from Earth, forming the 'right eye' of the so-called 'Man in the Moon'. The Imbrium Basin itself doesn't just contain the Mare Imbrium flow, but it is the largest one there. There are also two younger flows that are slightly smaller, yet still larger than any flows, historic or recent, we have on Earth. Scientists can sometimes identify these different lava flows from afar as they tend to be subtly distinct in colour. These differences, while small, help scientists to map out the extent of the flows and, after careful analytical and experimental laboratory work, they have concluded that the colour is related to the composition of the flow, with its varying amounts of iron and titanium.

If we want to look for comparable flows on Earth, then the largest basaltic eruptions we know about on our planet are the flood basalts, of which there are no modern examples. We touched on these flows in Chapter 2 in relation to supervolcanoes. And while we tend to think of supereruptions as the explosive type, such as Lake Toba, as we discussed previously, they are actually defined by the volume of their magmatic outpourings. As such, large basaltic eruptions such as the flood basalts that aren't necessarily as explosive can also fall into this category. While they don't form a huge caldera in a cataclysmic event, flood basalts can pump out large volumes of lava and associated gases in a geologically short timescale.

One of the best examples is the 16-million-year-old Columbia River Flood Basalt Group in the north-western USA. Yet, even here, some of the largest single flows are 'just' 750 kilometres (466 miles) long, which is only half that of the Mare Imbrium. Perhaps the Columbia River Flood Basalts would have formed an impressive mare on the Moon with its lower gravity? After all, the distance covered by this volume of lava is remarkable, and it is thought to have been erupted in just a few weeks.

It is, perhaps, easy to see why early astronomers might have thought the dark and vast regions of lunar maria represented bodies of water, but it wasn't only these apparent pools of rock that fooled astronomers into thinking the lunar surface hosted, or once hosted, flowing water. Intriguing features first observed in the eighteenth century and that look exactly like winding river channels exist within the basaltic flows. These enigmatic features are the sinuous rilles, with later investigation revealing them to be channels formed by fast-flowing, turbulent lava flows, as opposed to water. As the hot lava moved over the existing land, it is thought to have melted the rocks below while it carried on flowing, leaving behind a channel just like a dry riverbed. The largest of these channels is called Hadley Rille, and is more than 100 kilometres (60 miles) long, up to 3 kilometres (1.8 miles) wide and 1 kilometre (0.6 miles) deep. The Grand Canyon is only 1.8 kilometres (1.1 miles) deep – this puts into perspective how significant the rilles are. The Moon may be small, but its basaltic volcanic features are arguably more impressive than Earth's.

The Moon certainly doesn't have any large classic conical volcanoes like we see on Earth, which is understandable because of the nature of its runny basaltic flows and lack of gravity, which allows lava to spread out over great distances instead of building up to form elevated features. Nevertheless, the lunar surface does host some small cones and domes that resemble Earth-like volcanoes, but on a smaller scale than our large shield volcanoes. These cones and domes have steeper sides than basaltic shield volcanoes on Earth, being more similar in size and shape to the cinder cones and other small volcanic domes we find on our planet. The question is, how could these features have formed when the lunar lava is so runny? The difference for these lunar cones is that the lava forming them is thought to have been slightly cooler than the mare flows, representing the final stages of volcanic activity on the Moon. As volcanism wanes, because of a cooling planetary body, magmas become more viscous and, therefore, they are unable to flow very far from the vent, instead piling up around it. In some cases, these and similar features might even have been formed from more explosive pyroclastic deposits, just like a cinder cone on Earth. What's more, some of the lunar cones line up with each other, leading scientists to suggest they may have resulted from fissure-type eruptions similar to those seen at places such as Hawaii and Iceland, where lava essentially flows up through a crack in the ground as opposed to emanating from a single vent.

Despite lunar volcanism being dominated by basalt – a lava that tends to produce effusive rather than explosive eruptions – it doesn't mean that explosive eruptions didn't occur. The edges of the lunar maria are often surrounded by thousands of square kilometres of dark layers of materials that are most intriguing because they appear to have flowed uphill. These dark so-called 'mantling deposits' drape over the existing surface, forming a range of lumps and knobs and other high features. Despite the Moon's lower gravity, the laws of physics remain, and so lunar lavas should still only flow downhill. As such, the mantling deposits presented a conundrum. That was until they were sampled by Apollo 17 astronauts. They found these weird-looking deposits are made from small, orange and black spheres of glass. The so-called glass beads turned out to be superfast-cooled pieces of once liquid rock and are thought to have formed during eruptions that might have resembled Hawaiian fire fountains. The lunar variety were apparently more explosive than their earthly counterparts, because they are spread over much greater distances than we find for similar deposits on Earth. But let's not forget the gravity factor: of course, with lower gravity, debris can spread much further.

Whether the Moon saw explosive activity or not is important because the style of volcanism can tell scientists about what was happening on its insides over the course of history. However, as you'll see, this information could also play a key role in our future exploration of the Moon.

Scientists are certain that the magma erupted to produce the glass beads must have had a high volatile content, which was necessary to fuel the explosive nature of the eruptions. Even though these beads travelled a long way, the large distances cannot be accounted for by the Moon's lower gravity alone. The beads must have been ejected with great force thanks to dissolved volatiles that expanded as the magma reached the surface. But the problem has always been that for a long time the Moon was thought to be dry, inside and out, lacking the requisite water – dissolved inside along with other volatiles – that Earth and other places like it use to fuel explosive volcanic behaviour.

The reason scientists expected the Moon to be dry inside and out is because of its assumed formation during a cataclysmic impact, where temperatures would have exceeded that for which water could still be viable on or in the Moon. Nevertheless, in the last decade scientists have begun to use advanced laboratory techniques to measure the volatiles in lunar samples, finding that they contain much more water than they ever expected. They have now concluded that the Moon is not completely depleted in volatile elements such as water.

Understanding why this is the case is another question entirely. New models are required to work out how the Moon might have been able to hold on to its water during its violent and hot formation. It might be that such volatile elements were delivered later – after the Moon had cooled down – during comet and/or asteroid impacts. But either way, the fact that water has been detected within a wide range of lunar rocks, and even found within ice deposits at the lunar poles, means that scientists can conclude that the interior of the Moon – its mantle – contains water. This water, and other volatiles such as sulphur, chlorine and carbon monoxide, to name but a few, can account for the extra 'fuel' that is needed to kick-start those fire-fountain-type explosive eruptions.

The fact that the Moon is now known to contain water opens some exciting opportunities for our future exploration of space. When we eventually send humans back to the Moon and even set up a permanent base for exploration of the lunar surface and surrounding Solar System, it will be incredibly useful if they are able to obtain their own water in situ. The main alternative option is to bring water from Earth, but carrying water to the Moon is costly due to its volume and weight. To maintain a permanent human base on the Moon without water in situ would , therefore, be unsustainable. If we can extract the water from Moon rocks to biologically support astronauts over time then this is a much better approach. This same water could even be used as rocket fuel, if broken down into its constituent elements of hydrogen and oxygen, meaning we don't need to transport fuel from the Earth to the Moon, making it easier for us to explore well beyond the Moon. This may all sound like science fiction, but scientists working in the field of In Situ Resource Utilization (ISRU) have been investigating at how to extract water from Moon rocks for several years and it is proving to be a viable technique. We might wonder if Moon water will taste any different from Earth water. Hopefully one day soon some of us will get the opportunity to find out.

The lack of recent volcanism on the Moon might lead you to expect its insides today are cold and solid, just like the surface. But this assumption might not be correct. While the Moon has, seemingly, long since passed the point at which its interior is warm enough to produce rising magmas that can erupt at the surface, it might not be completely cold and solid. It's a small body, so we would fully expect it to have lost its primordial heat, yet studies suggest there might still be liquid material deep within the lunar interior. The Moon's core is thought to be around 1,400°C (2,552°F), which is not cold by any stretch of the imagination. Studies relying on moonquake data from the seismometers placed by Apollo astronauts have mapped out the interior structure of the Moon and revealed that it likely has a liquid outer core, just like our own. However, the seismic data is not as 'neat' as that for our own planet because the seismometers use technology from the 1960s, there are fewer of them, and the duration of the observation is much shorter than accumulated on Earth. This is coupled with the fact that moonquakes are not as frequent or as powerful as earthquakes, which means scientists do not have enough data to be certain what is down there. In fact, the outer core might more accurately be described as 'soft', but the implication is that it is, at the least, partially molten or mushy.

There is further evidence that the region at the very base of the lunar mantle, next to the core, might also be molten. If this is the case, then we might wonder why this molten rock hasn't risen to the surface. It could be that it is simply trapped by solid rock above. Or it might be that this deep, potentially molten rock is unable to rise because it is of a denser composition than the rock which surrounds it. Some scientists have suggested that this portion of the lunar mantle could be made up of titanium–rich rock that formed at the surface of the Moon before sinking down into the deep interior when the Moon was a global magma ocean, because of its higher density than the surrounding magma. For now, whatever this layer turns out to be, it appears to be stuck, possibly for ever.

It might seem that, apart from the occasional fire fountain, the only volcanism on the Moon is of the flowy, mare type. However, other intriguing mountains spotted on the lunar surface almost certainly require a different explanation. These landforms were first known as 'red spots' because of their red appearance as viewed by spectrographs – scientific instruments that break down light into its component parts to study characteristics like temperature, density and chemical composition. The space instruments that looked at these domes of rock found them to be as large as 20 kilometres (12 miles) across and 1 to 2 kilometres (0.6 to 1.2 miles) tall. It is impossible for mare-type basalt to have piled up to form mounds of this size because it is simply too runny, and so these landforms caused some confusion when they were first observed. After further study it was concluded that the domes were composed of materials such as the mineral quartz, which is very rich in silica, silica-rich glass and alkali feldspar. These are not constituents of basalt, but instead represent silicic volcanism.

The silicic mineralogy that was observed actually made a lot of sense because the shape and size of the red spots made it obvious that they must have formed from lava that was much more viscous than standard basalt. The lunar red spots are reminiscent of Earth's stratovolcanoes, those we find at subduction zones. But because of the lack of plate tectonics on the Moon, the steep-sided lunar domes cannot have been produced in the same way. Scientists noted that these highly silicic domes are located close to some of the mare basalts, which, intriguingly, is a relationship often seen on Earth – with more silicic lavas found in close proximity, and seemingly related to, runny basaltic volcanism.

On Earth, we find that these more silicic, sticky magmas can form in two ways. One is when a runny basaltic magma sits in a magma chamber under the surface for some time and evolves, fractionating, building up gases, and turning into a more silicic, sticky magma. The other is when the heat of a basaltic magma melts the old crustal materials within which it sits, incorporating them into the liquid magma and changing its composition to one that is stickier and silica-rich. For the time being, it is unknown which, if either, of these processes might have formed the Moon's red spots, and so they remain an intriguing feature. Either way, from the detailed study of some of these domes, it seems that basaltic and silicic lavas were erupted simultaneously, so clearly the process that produced these landforms is complex.

One of the aspects we haven't discussed is whether lunar volcanoes might still erupt. We've seen that much of the Moon's volcanism is ancient and that it was very active early in its history. But today, the Moon is often described as cold and dead, so it might come as a surprise to learn that scientists don't think the Moon's volcanic activity is necessarily over. NASA's Lunar Reconnaissance Orbiter has been scanning the Moon since 2009 at a much higher resolution than has been possible previously and it has found many structures that have never been seen before. Some of these appear to be lava flows, and based on how they contact the surrounding rock and the lack of craters on them, scientists suggest these flows must be relatively young. While the mare basalts were active around three billion years ago and almost all petered out by about one billion years ago, such evidence suggests the Moon experienced some more recent bursts of activity, possibly within the last 10–100 million years. And it might not be over yet. It's unlikely that we'll see any activity within our lifetimes, but if we were to, the lava is not expected to be of the large, runny mare-type but instead much more viscous, probably with a consistency more similar to thick soup. Any future eruption for the Moon is likely to be completely sporadic and short-lived; the Moon's volcanic history truly has petered out to almost nothing.

But still, you might be wondering how the Moon could show any burst of activity at all in the future if it's so dead at the surface now? The answer, as we learnt earlier, is that the Moon's insides are still warm. While they are not warm enough to fuel constant volcanic activity at the lunar surface, the heat within the Moon still needs to find a way to escape into space. It loses some of this latent heat gradually, seeping out slowly by conduction, but there is a chance that there might be just enough heat to move some molten rock around somewhere within its interior. Any future activity really would produce a burst of an eruption, because as we've seen, the lava will be sticky. Watch this space!

## The active one: Io

Io is a moon orbiting Jupiter that is not dissimilar in size to our own Moon. Therefore, before the first exploration with spacecraft, it was expected to be cold and dead because of its size. Yet, as we've seen, when Voyager 1 beamed back images of this alien world for the first time in 1979, it was soon apparent that the rocky lands of Io were being actively transformed with new lava flows, literally as the spacecraft flew past.

What the images returned by Voyager 1 showed were features such as the Ra Patera volcano on which, despite the poor quality of the raw images, scientists could see a dark caldera-like blob with what could only be described initially as 'some smudges' around the edges. These smudges, it turns out, were lava flows. While this was a surprise to many – and a very exciting discovery to learn that Earth was not the only volcanically active planet in the

Solar System – scientists had, in fact, already hinted at this possibility in the preceding years. Earlier in the 1970s, astronomical observations revealed Io's surface was covered in sulphur, and that it lacked water ice. No one could find a reasonable explanation for where the sulphur might have come from apart from the suggestion it might have been produced on Io by a volcano. While this conclusion was met with great scepticism at the time, as people didn't expect Io to be active, later in the 1970s there was further evidence in support of the idea.

In 1979, but still prior to the Voyager encounter, telescope measurements detected a brightening in some of the infrared wavelengths (radiant light that is invisible to the human eye but that we feel as heat) on Io. The bright spots were deemed to be hotter regions on Io's surface, and scientists were able to estimate temperatures for them of over 400°C (752°F), much hotter than Io's normal daytime surface temperatures. Active volcanism seemed like the most plausible explanation for these differences because there didn't seem to be any other way to account for such temperatures. Yet the scientists were still sceptical because they didn't understand how Io could be fuelling lava flows when it was so small, and so far from the Sun.

As we saw briefly in Chapter 7, this is where things got even more interesting when, later in 1979, just two weeks before the Voyager 1 encounter with Io, a team of scientists predicted something that was to prove very important. Their work suggested that Io's interior was squashed and stretched by the gravity of Jupiter, coupled with the tug of the other Galilean moons, which acted to knead the inside of Io as they went about their orbits. Such tidal heating remains the best way to explain how the insides of Io are heated to tremendous temperatures thanks to the friction of its internal rocks rubbing against each other, producing the now famous volcanic activity at the surface. The implication of this, however, was that the whole of Io could be expected to be considerably volcanically active. At the time, this still seemed like a big stretch of the imagination for such a small moon.

As a result, when Voyager beamed back its revealing images, they didn't disappoint. What was uncovered was a 'pizza' moon, covered in spots that looked like pepperoni on a pizza. The 'pepperoni' pieces weren't impact craters made by comets and asteroids, as you might have expected them to be on a long-dead world, but instead volcanic craters. Even though none of the volcanoes that held these craters were seen to be actively erupting at the time, the fact that they *weren't* impact craters strongly supported the idea that Io was an active world. Yet scientists were missing the smoking gun to prove for certain that this was a volcanic world: seeing volcanoes erupting in real time. Thankfully, it wasn't long before they got their proof.

A few days after Voyager's encounter, strange cloud-like features, hundreds of kilometres above the surface, were spotted on some of the images that had been returned. These features were found to coincide geographically with regions of Io that were hotter, the first with what we now know as the Pele volcano, and there was only one explanation. The clouds were, in fact, plumes associated with hotspots of volcanic activity on Io's surface. Pele wasn't just a volcano, but an active one.

Since the 1970s, Io has been visited by the Voyager 2 spacecraft, four months after Voyager 1, and then Galileo, which entered orbit around Jupiter and made observations of the planet and its moons. New Horizons even flew by on its way to Pluto in 2007, and NASA's Juno spacecraft, launched in 2011, is in orbit around Jupiter, obtaining more information about the planetary system and its moons. What has been revealed by these visits is that Io is the most volcanically active object in the Solar System, with the hottest lavas and the largest eruptions. Io is truly a world of fire and ice. Near to its volcanoes, temperatures can be 1,650°C (3,002°F), but in other regions average temperatures can be as low as -130°C (-202°F), where fields of frozen sulphur dioxide form. Io can get down to such cold temperatures because it is sometimes in the shadow of the giant planet Jupiter, which totally blocks out the Sun's rays, but it is always hot inside.

When Voyager 2 visited Io, it imaged the Pele volcano that had earlier been imaged by Voyager 1 with a plume above. Even though only four months had passed, distinct changes to the region were visible. The active plume over Pele was no longer present, but there were two new ones, and the shape of the volcano on the ground had changed from what was originally described as heart-shaped into a more oval shape. These differences were significant, with the area of change estimated to be around 10,000 square kilometres (3,800 square miles). Pele was indeed a very active volcano, and it wasn't alone.

The Galileo spacecraft, named after Galileo Galilei, was launched in 1989. However, it didn't start to study Io until 1996 and even then, it had to keep its distance because of the extremely harsh radiation environment in the region of Jupiter. The spacecraft was to experience 4,000 times the human lethal dose of radiation thanks to Jupiter's very powerful magnetic field, which isn't good for spacecraft electronics either.

Before the Galileo spacecraft began to study Io, there was much debate as to what its volcanoes were producing. Scientists needed to be able to account for all the sulphur on Io's surface. When we see images of Io, it is a stunning mix of oranges, reds and yellows, like an artist's mixing palette. These swathes of colour were determined to be rich in sulphur and scientists had an idea that it might be produced as sulphurous lava flows emanating from the volcanoes. This would mean that Io was quite unlike its terrestrial friends in the inner Solar System in not erupting silicate rock. Still, Io had already given us quite a few surprises, so anything seemed possible.

Around this time, the temperatures recorded for the volcanic eruptions on Io didn't seem to be high enough for them to be made of molten silicate rock, which requires high temperatures to melt, at well over 600°C (1,112°F). Temperatures lower than this, like those measured on Io, were a better match for molten sulphur, which has a lower melting point. Sulphur is a strange substance because it boils above 444.6°C (832.3°F), but can remain molten down to 115.2°C (239.4°F), meaning that lava flows made of sulphur could be flowing at relatively cool temperatures. The especially weird thing about a sulphur lava flow is that it would become much less viscous, not more, as it cooled to around 175°C (347°F), the opposite of what you would expect for a silicate lava flow. Clearly, it would be good to understand whether Io has these strange flows.

Despite the evidence, scientists weren't convinced Io was erupting molten sulphur, so they continued to study it given that relatively little was known about this tiny moon of Jupiter. Fortunately, astronomers were able to make some detections remotely from Earth, and what they found was much higher temperatures than had been seen before. These temperatures extended above the range for molten silicate rock eruptions of over 700°C (1,292°F), much higher than the boiling point of sulphur. A camera system on Galileo corroborated this; it found small regions on Io's surface with temperatures of 725°C (1,337°F) and higher, which were interpreted to be either lava flows or actively convecting lava lakes.

Lava lakes are relatively uncommon on Earth, but they are now known to be common across Io. They are exactly what they sound like: a lake consisting of lava, usually contained within a volcanic vent or caldera. The most famous example on Earth in recent years is Kilauea on the island of Hawai'i, which actually has two lava lakes, both of which dramatically emptied in 2018 because of increased activity at the volcano before filling up again with lava. However, the most terrifying example we looked at was the lava lake contained at the top of Nyiragongo volcano, in the Virunga National Park, Democratic Republic of the Congo, in 1977. When the caldera walls that were holding the lava lake in place fractured, they released the entire contents in an hour.

Lava lakes tend to refill over time, as magma below the volcano summit or caldera continues to be produced, and so where we have them, they tend to be long-lived features, on human timescales anyway, lasting for at least a few decades. On Io, lava lakes seem to be similarly long-lived. Voyager observed one at Loki Patera, Io's largest volcanic depression, which is over 200 kilometres (120 miles) in diameter. It was still there when Galileo was in orbit. But this is a supersized lava lake when we compare it to any we know on Earth, encompassing an area a million times larger than a typical one here. Galileo also observed some interesting features associated with this lava lake. The edges near the western caldera walls are hotter. Combining these observations with ground-based data from telescopes, scientists were able to conclude that Loki's lava lake periodically overturns. This is because sitting atop the molten reservoir of lava within the caldera is a thin, solidified lava crust. Every so often, because this fragile crust is denser than the molten lava, it sinks back in, in an event that signals an unstable phase for the volcano with a huge release of heat. The hot edges that were observed are where fresh lava floods up to fill the gap left behind by the sunken crust. Gradually the thin crust starts to form again: the start of a new, quieter phase for the volcano. Scientists have observed these events and have found they occur roughly every 500 days.

Io has the hottest lavas of any place in the Solar System, with temperatures commonly exceeding 1,500°C (2,732°F), with one lava possibly even reaching above 1,700°C (3,092°F). This is much hotter than Earth's volcanoes but is actually more similar to the temperatures that scientists have calculated for lavas that predominantly flowed here billions of years ago. Such lavas are known as ultramafic, characterised by very low silica contents (below 45 per cent), with higher amounts of magnesium. In some ways they are not dissimilar to basalts, but their slightly different chemical properties make them even more runny. If Io's lavas are ultramafic in composition, then they might provide a window into understanding Earth's ancient volcanic past.

Whether Io erupts ultramafic or basaltic (otherwise known as mafic) lavas is an interesting scientific question and one that will continue to be investigated. In particular, NASA's Europa Clipper mission and ESA's JUICE (Jupiter Icy Moons Explorer) will be studying Io along with Jupiter within the next few decades. But while we've learnt that Io's lava flows are made of basalt, or something very similar to it, we haven't answered the question of where all the sulphur comes from.

It turns out sulphur is produced by Io's volcanoes, but not as lava flows. Instead, it is a secondary product that is either one of the main volatiles to be pumped out during eruptions in plumes, or is incorporated into eruptions when vaporised from the sulphur dioxide frost on the surface as an eruption continues. On Earth, the main volatiles released by volcanoes are usually water and carbon dioxide, but there are some examples of huge volcanic releases of sulphur dioxide into our atmosphere too: namely, the 1783–4 eruption of Laki in Iceland and the 1815 eruption of Mount Tambora in the Philippines. As we've seen in Chapters 1 and 2, these eruptions both had significant, yet short-lived, effects on the Earth's climate, with major atmospheric cooling events ensuing soon after the eruptions. Io doesn't pump out water and carbon dioxide, but it still releases lots of sulphur and sulphur dioxide, along with more minor amounts of sodium, potassium and chlorine. Sulphur is relatively dense and not very volatile compared to other gases, and because it is erupted into a very cold environment on Io, it forms 'volcanic snow'. This is precisely what it sounds like: condensed and frozen pieces of sulphur dioxide that have been ejected by a volcano, and then fall back to the surface to form frost.

Despite Io being a very active world, most of its surface was not, in fact, resurfaced as a result of lava flows, because they cover a relatively limited area. Rather, its surface is mostly covered by the products of its plumes. The plumes that release sulphur dioxide on Io also release huge amounts of dust, and they come in a few different varieties. There are some that are long-lived, known as Prometheus, with four examples of these observed by Voyager 1 in 1979, Galileo in the late 1990s and the New Horizons mission in 2007 at the same location, decades apart. Prometheus plumes tend to be no taller than 100 kilometres (60 miles) and are produced when an encroaching lava flow meets some sulphur dioxide frost, causing it to vaporise and shoot material skywards. These plumes tend to be very dusty when the lava is blown apart on contact with the volatile material. Such plumes can be likened in some ways to the Old Faithful geyser in Yellowstone National Park because they are constantly erupting. But, unlike Old Faithful, they are rocky magma and not water, and so have long lava flows associated with them. The Prometheus volcano, which gave its name to this type of plume, has lava flows over 100 kilometres (60 miles) in length.

If it is explosive eruptions you like, then you want to look at the Pele plumes. These are so powerful they can reach heights of up to 500 kilometres (310 miles), somewhat higher than the orbit of the International Space Station above the Earth! They occur when sulphur and sulphur dioxide exsolve from magma at volcanic vents or in lava lakes. In August 2001, a pass of Galileo across Io was unknowingly timed well to catch, and sniff, one of these plumes during a gigantic eruption of Io's Thor volcano. Galileo's plasma science experiment detected sulphur dioxide gas and very fine dust grains within the plume and when it passed over again later that year it found a vigorously erupting volcano with images revealing new dark lava flows.

Even from distant observations scientists can see that Io hosts hundreds of volcanoes and hundreds more hotspots representing other regions of active volcanism. This fascinating world, which is constantly being heaved and pulled around by Jupiter and its neighbouring moons, while different from Earth in many ways, has the potential to teach us a great deal about our planet's early history.

## Asteroid fire

While we've focused here on the small, rocky active moons of our Solar System, we mustn't ignore the even smaller objects, the asteroids. We've seen again and again that to produce a volcanically active world, we need a body large enough to retain heat, so it might seem surprising that an asteroid is thought to be capable of producing a volcano.

And while the evidence for an asteroid volcano has not yet been found, scientists think there is every possibility that they might have erupted in the past. But here we would be talking about very ancient eruptions in the Solar System's early days, around 4.5 billion years ago.

The asteroids are mostly small rocky and metallic objects that represent the building blocks of the planets, the leftover remnants from the time of planet formation. They experienced similar events to the planets, with many being large and hot enough to undergo internal differentiation, where the metal within them sunk to the centre to form a core surrounded by a silicate mantle and crust, just like the terrestrial planetary bodies. But like the planets, the asteroids have also gone through a complex history of collisions, the difference being that they were much more likely to be broken apart because of their small size. It is during this process that asteroidal volcanoes may have erupted.

Early on, when the asteroids were still hot worlds, with molten metallic cores that were sinking away from the rock within them, large collisions would have had the effect of ripping off the silicate mantle, which was probably cooler, and exposing the molten metallic cores. Sometimes these too were disrupted. All of this rocky and metallic material would be flying about the Solar System and can account for many of the iron-rich and rocky meteorites we have in our collections on Earth today, as they have gradually found their way here over millennia. But before some of these newly disrupted blobs of free-floating iron cores cooled down completely, particularly if they were large, they may have themselves produced iron-volcanoes, otherwise known as ferro-volcanoes. As the outside of the metallic blobs cooled, the insides would have remained insulated and thus molten for longer and, as we know, a liquid is less dense than its solid, so will have more buoyancy. As such, liquid iron may have worked its way up to the surface to erupt out on the asteroid's infant surface.

While this might seem a bit far-fetched – and very hard to test because it took place billions of years ago, if at all – we might get the chance to find out sooner than you would expect. NASA is planning a mission, called Psyche, that will make its way to asteroid Psyche 16 in the asteroid belt. Psyche 16 is an iron-rich object that has thrown up some interesting questions, partly because it has a lower density than its metallic surface would suggest. Could it be that Psyche 16 has a molten iron core that erupted up through a thin silicate mantle and crust long ago to cover its surface in iron? We will have to keep our eyes on this mission for more information but, if so, then we have a brand-new type of Solar System volcano to discuss.

CHAPTER TEN
# Fiery Planets

## The little one: Mercury

Mercury is the Solar System's smallest planet, just a little larger than our Moon, and it also happens to be the closest to the Sun. Despite its planetary status, its surface looks very similar to the Moon's: it is heavily cratered, indicating inactivity for billions of years. That's not where the similarities end though. Mercury also shares the Moon's mix of landscapes, first observed by NASA's Mariner 10: crater-covered highlands, thought to be very ancient indeed, and extensive smooth plains, thought to be more recent deposits, geologically speaking, such as lava flows. Mercury also has some impressive valleys, which look very much like those carved by rivers on Earth. However, liquid water is not stable at Mercury's surface pressure and temperature, so these valleys must have been carved by something else. That something else was probably lava. Nevertheless, there does appear to be water ice within Mercury's permanently shadowed polar craters, which presents a bit of a paradox because the planet reaches such high daytime temperatures.

But Mercury gets even more fascinating because part of the reason it is so small is because it's composed largely of core. In fact, Mercury's core makes up 55 per cent of its volume, which, compared to Earth, whose core is just 17 per cent of its volume, is huge. This means Mercury has much less mantle and crust than Earth, with perhaps as little as 400 kilometres (250 miles), and certainly no more than 700 kilometres (430 miles), of silicate surrounding its core, compared to Earth's 2,800 kilometres (1,700 miles).

To date, Mercury has been visited by two spacecraft: the three Mariner 10 fly-bys between 1975 and 1976, and MESSENGER from 2008 to 2015, with no landers or humans exploring the surface. Fortunately, the BepiColombo spacecraft will reach Mercury in 2025, encompassing two orbiters that will carry out a comprehensive study of the planet to learn about its interior and exterior structure, which should reveal a wealth of important information about the planet's history. Despite the limited studies of Mercury to date, we do know quite a bit already and look forward to adding to that knowledge soon.

Mercury's surface is thought to be covered by extensive lava flows, which tend to be concentrated in the northern hemisphere. They present as smooth plains that occupy around 27 per cent of the planet's surface, made up of young flows and older, more heavily cratered flows. Reaching this conclusion about Mercury was not easy because the images returned from Mariner 10 led to some confusion. The smooth plains appeared lighter and brighter than the rocks surrounding them, contrary to the lunar mare, which are darker than their surrounding rocks. Nevertheless, when scientists calculated the absolute albedo (the brightness) of the smooth plains, they were found to be the same as the lunar maria. The reason they looked lighter is just because Mercury is dark everywhere, making those areas appear very bright in comparison. Some scientists used the similarity in absolute albedo as evidence that Mercury's smooth plains could also be basalts, like the lunar mare.

Despite this evidence, it still wasn't clear if Mercury's plains could instead be broken rocks, known as breccia, created during impacts onto the surface. Fortunately, NASA's MESSENGER orbiter studied these regions in much more detail, with the high resolution images returned revealing that the plains appeared to have been made from lavas that had flooded the landscape, filling in depressions and appearing very much like the classic flood basalt lava flows we find so commonly on Earth and other planetary bodies, including the Moon.

One of these regions, informally named the Northern Smooth Plains but more formally Borealis Planitia, was spotted in the northern polar region of Mercury. It is notable because of its size, which dwarfs pretty much all known examples of flood basalts here on Earth, covering 7 per cent of the planet; it is up to 1.8 kilometres (1.1 miles) thick, with an estimated volume of around 2 to 3 million cubic kilometres (720,000 cubic miles). Scientists concluded the Northern Smooth Plains resulted from a single phase of volcanism, albeit one that might have lasted a very long time – possibly several tens of millions of years – in order to have produced such a large volume of lava.

The flows that formed the smooth plains were probably also responsible for carving out Mercury's valleys, or 'valles' as they are known. The valles are characterised by smooth floors and steep sides that can be up to 20 kilometres (12 miles) wide and very long, stretching to tens or even hundreds of kilometres. The valles are thought to have been created when the heat of the lava flows incised the ground, acting to erode it into sharp valleys. However, whether this occurred during one or many lava flows travelling down each valle is unknown. It has been noted that, in some cases, not all the lava that flowed down the valles was confined to the channel, as, in many places, the images show lava that has flooded out over the surrounding landscape, like a river breaking its banks, probably during periods when the lava eruption rate was just too high to be contained and thus inundated the valley.

Just like the Moon, Mercury has had its fair share of explosive volcanism and it is, in fact, something the Mariner 10 scientists specifically looked for, because of the known occurrence of explosive volcanism associated with the waning of flood basalt eruptions on Earth. For explosive volcanism on Mercury to occur, there would need to be volatiles present, like water and carbon dioxide. However, these molecules are not expected to be stable on Mercury because, being volatile, they are easily lost at Mercury's high surface temperature. It was, therefore, quite a big surprise when MESSENGER detected volatiles on Mercury, such as sulphur, carbon, potassium and chlorine, and not just that they were there, but that they were there in large quantities. Even if we can't explain why they exist on the closest planet to the Sun, their presence helps explain how Mercury could have experienced explosive volcanism, with the volatiles helping to drive powerful eruptions.

In relation to this, MESSENGER encountered over 100 so-called 'bright red spots', now formally designated 'faculae', scattered over Mercury's surface. These features were commonly around 10 to 50 kilometres (6 to 30 miles) wide, but the largest example has a recorded diameter of 270 kilometres (170 miles). Scientists interpreted these to be volcanic vents, with the spots surrounding the holes being the deposits of material thrown out by explosive eruptions from the vents. Such 'pyroclastic vents' across Mercury are thought to have been produced during the latter stages of flood basalt volcanism, just like on the Moon and Earth.

Looking at the timing of Mercury's volcanism, we have a problem, in that we have no physical rock samples of any of Mercury's lava flows, which would enable us measure their absolute ages.* In fact, we have no samples of Mercury at all. Surprisingly, out of all the meteorites we have on Earth, we don't think any originated as rocks on Mercury. Furthermore, since we haven't launched a sample return mission to the innermost planet, we are seriously lacking any 'ground truth data'. Nevertheless, what scientists can do is use crater-size frequency distribution investigations (sometimes known simply as 'crater counting') to estimate the age of different regions of terrain. It might sound inexact, but they can compare the crater distributions on Mercury with those determined on the Moon and then combine this with our knowledge of the age of lunar terrains from the rock samples the Apollo astronauts brought back. Scientists can then use this method to estimate the absolute ages of regions of Mercury's surface.

It turns out that crater counting does a good job. Crater-size frequency analyses reveal that plains, such as the Northern Smooth Plains, were formed around 3.7 billion years ago and that there was no major volcanism on Mercury after about 3.5 billion years ago. However, it might be that some of the more explosive activity extended into the last billion years of the planet's history, possibly as recently as around 300 million years ago. Unfortunately, this is hard to confirm at present because the red spots that mark out the explosive activity are very small, which makes crater counting less accurate (fewer data leads to worse statistics). Either way, what we can conclude from this is that Mercury's purported history contrasts with that of Earth, Venus and Mars, which all experienced long multi-billion-year volcanic activity, of which we'll

---

* An absolute age is determined using radiometric dating and gives a value in Earth years of the age of the rock, as opposed to a relative age in relation to the rocks surrounding it.

learn more soon. However, it is in keeping with that experienced by the Moon, a planetary body of a similar size.

Because it has been a long time since Mercury saw any volcanic activity – even if geologists class 300 million years ago as 'relatively recent' – a surprising feature of the planet is its magnetic field, which, although weak, suggests that there must be some residual heat within the planet. Mercury is the only planet in the inner Solar System, other than Earth, to host a magnetic field. As we've seen with Earth, a magnetic field supports the idea of a churning molten core, and ours is vital in creating a safe environment to host life. Yet, on the surface, Mercury appears to be as dead as the Moon. Just like the Moon, Mercury's small size means we would have expected it to have cooled down internally by now, despite its surface being very hot at times because it orbits so close to the Sun. But the presence of Mercury's magnetic field serves to highlight how different two celestial bodies can be deep inside, even if they appear similar on the surface.

Although the heat within Mercury is not enough to fuel volcanoes, the planet still experiences tectonic activity. This is thanks to the fact that as Mercury's hot, dense, iron-rich core continues to cool, it is solidifying, and metal sinks towards the centre of the planet to form an inner solid core. As solids cool, they tend to contract, meaning that the planet shrinks. As this process has continued over the last few billions of years, the contraction has caused Mercury's surface to crumple, creating the scarps and cliffs and other big folds that criss-cross its surface, cutting across craters and proving that they are, therefore, a more recent geological feature. You might not think these tectonic events are very significant, but Mercury's 'Great Valley' is around 1,000 kilometres (620 miles) long and 3.2 kilometres (2 miles) deep, making it larger than the Grand Canyon.

Scientists have suggested that over the course of Mercury's contraction, it has shrunk by at least 7 kilometres (4.5 miles) in radius, which is certainly enough to account for the features observed. And it's not over yet because Mercury is still cooling and, as it does so, its tectonic activity continues. Mercury may yet rumble with more earthquakes, or 'mercuryquakes'.

## The twin: Venus

We've already met our closest planetary neighbour, Venus, a few times. As we've seen, it is the most similar to Earth in terms of size, and it has a similar mass and radius. Venus was formed from the same chemical building blocks as Earth, in roughly the same part of the Solar System. Like our own planet, Venus is covered in lava flows and it also has very few impact craters scattered across its surface. The lack of dense cratering indicates that Venus' lava flows could have flowed relatively recently, such that the craters produced by comets and asteroids bombarding the planet over the course of history were covered over and buried by the lava, and hence lost within the geological record. However, unlike the volcanoes on Io, and despite Venus having more volcanoes than any other planet in the Solar System, we've never yet seen one erupt. As such, it is very hard to know if Venus is still active today, or if it has been in the relatively recent geological past. Furthermore, as we saw in Chapter 5, scientists may have detected phosphine within the atmosphere of Venus, a gas that they suggest could have been produced biotically. Although its discovery and the subsequent interpretation of its formation are all very much under debate, despite first appearances, Venus may be a life-giving planet.

Looking at its volcanic history, we don't know exactly how many volcanoes Venus has in total, partly because there are so many of them, perhaps over a million, with an estimated 1,600 major ones. The big conundrum, as we've seen, is whether any of them are active today, or have been in the recent past. The major problem in assessing this is that we have a hard time seeing the Venusian surface. Venus is always shrouded in a thick haze of carbon dioxide, the main gas making up its dense and toxic atmosphere. Studying its surface from orbit requires radar, otherwise we are unable to see through the impenetrable haze. Fortunately, we've sent spacecraft to Venus equipped with radars to reveal the surface in detail. Whereas visible light gets reflected, or scattered away, by Venus' clouds, radar can 'see' the surface thanks to its longer wavelengths, which means it can penetrate through clouds to reach the surface. During the 1990s the Magellan spacecraft mapped nearly the entire surface of Venus using radar. But despite revealing an intricately detailed surface, it was still not a simple task to identify the features that radar showed. The reason is that bright spots in radar maps can be due to a change of elevation, a difference in rock composition or changes in the physical properties of the surface such as texture, porosity and/or density. This meant that the best way for scientists to figure out what the radar maps were showing was for them to assemble the Venusian images and compare the features they showed to images collected by similar studies on our own planet. Here we know what the features are as we have the ground truth, so we can apply this knowledge to Venus.

Thanks to this work, scientists have discovered that lava flows from thousands upon thousands of volcanoes have shaped Venus' surface, with more than 90 per cent of the planet covered by basalt. Yet the Magellan data revealed a surprising amount of variety within these flows: from vast outpourings of floods of lava, to steep-sided volcanic vents and even channels carved by lava. What was revealed was a Hawaiian-like landscape.

Focusing on Venus' largest volcanoes, which are few and far between, they don't appear to be aligned in a way that would indicate plate tectonics has been a feature of the planet. Venus almost certainly has only one plate: it's a so-called 'stagnant lid' body, just like that on Mars and Mercury. Venus' largest volcano is Maat Mons, an enormous shield volcano not unlike those seen in Hawaii or at other hotspot locations, and very similar in size. Maat Mons sits at around 8 kilometres (5 miles or 26,000 feet) high, only a bit shorter than Hawai'i's Mauna Loa, which is just over 9 kilometres when measured from the seafloor (5.6 miles or 30,000 feet). But this is nowhere near as high as Mars' Olympus Mons, which is almost three times taller. This might seem strange because without plate tectonics we know that mantle plume volcanoes are able to keep growing in one spot and can become very large, like those on Mars. In contrast, Earth's shield volcanoes fail to grow very large because plate tectonics moves the plate away from the mantle plume gradually over time. However, despite Venus also having a stagnant lid like Mars, Venus' volcanoes have been unable to grow to the same lofty heights. Why might this be?

To make a big volcano, you need a solid surface upon which it can grow. And this, in part, is why Venus lacks them: its crust is too squishy. Venus' soft, Play-Doh-like surface is a result of its greenhouse atmosphere, which creates a ridiculously hot surface, sitting at around 450°C (842°F), temperatures hot enough to melt lead. But Venus' dense atmosphere is also to blame as it creates a high surface pressure, more than 90 times that of Earth, another factor that prevents mountains from growing very tall. Therefore, when lava flows out from Venus' volcanoes, it tends to travel a long distance because it is very fluid and can move a long way over gentle slopes, unlike a thicker and more viscous lava, which would halt closer to the volcano.

Over the course of the space age, scientists studying Venus have concluded that the first 80 per cent of its history is no longer exposed at the surface, meaning that the rocks that were produced during this time have been lost or covered over with younger rocks erupted more recently. This means that the surface we see today on Venus is all roughly the same age, geologically speaking, which is, give or take, around 500 million years old. This is in stark contrast to Earth, which preserves surface rocks dating back almost all the way to our planet's formation 4.5 billion years ago, despite the fact that we have plate tectonics, which has destroyed much of our history. Earth has rocks that vary from just over 4.2 billion years old, through all the time periods in between then and now, which allows us to piece together a detailed and protracted history of our planet.

The relatively uniform age of Venus' crust suggests that the planet must have experienced a very significant geological event in the most recent third of its history. Going by the fact that most of Venus is thought to be covered in lava flows, we can guess that the culprit was volcanic activity. But the lavas that flowed during that time, somewhere between 300 and 600 million years ago, must have been large enough to resurface much of Venus in a relatively short timescale, again geologically speaking. If we imagine the same event on Earth, then we would be looking at covering an area encompassing all our planet's oceans and land masses, except for the Pacific Ocean, with fresh lava flows within just a few hundred million years.

Flood basalts are the most obvious activity, but if they are to account for this huge amount of lava, then they are like none we've ever seen on Earth. Scientists aren't sure precisely why Venus went through such a major resurfacing event, but it might be related to its 'stagnant lid' and lack of plate tectonics. Venus' mantle convects just as ours does on Earth. But, with a lithosphere above that doesn't move, the heat can only travel through it by conduction, which is not very efficient, so the internal heat continues to accumulate within the planet. On Earth, we have useful gaps at the edges of our tectonic plates that are efficient at letting heat escape. But on Venus, over time, it is thought that its upper mantle builds up heat. Around 600 million years ago this is presumed to have reached the point where the temperature was high enough within Venus for large-scale melting to occur within its upper mantle. When this happened, Venus could have become mechanically unstable: solid, dense lithosphere finds itself sitting atop a molten, more buoyant asthenosphere. In such a scenario, the whole system is thought to be capable of 'overturning' in a relatively short geological timescale, say over 100 million years, resulting in a period of intense volcanic activity that releases all the pent-up heat as lava flows. In a way, this is a little bit like a brief period of plate tectonics, as the solid lithosphere sinks into the molten layer below, which is rising up to fuel extensive lava flows at the surface. It is also similar to the overturn of Io's lava lake that we met in the previous chapter. The main problem with such an event from a scientist's point of view is that it will have erased the past geological record, removing any hope of finding evidence for the existence of plate tectonics on Venus, if it did occur in the past. Not only that, but not being able to see or sample rocks older than around 500 million years on Venus means that we will struggle to piece together its early years and, in turn, find out if it ever hosted a suitable environment for life. Despite its toxic atmosphere now, it might not have always been this way.

It seems that Venus was very active around 300 to 600 million years ago, but what does that mean for volcanic activity since then? From extensive mapping of Venus from orbit, scientists have concluded that there was a major drop-off in volcanic activity around 300 million years ago, which is mostly based on cratering history. When looking at Venus from afar, it is no easy task to figure out when any lava flowed, so assigning an exact age to any region of Venus' crust is hard. Nevertheless, small changes in Venus' surface temperature have been observed with spacecraft instruments that look at variations in visible and infrared light. Instruments on ESA's Venus Express Orbiter measured the amount of infrared light emitted from Venus' surface during night-time. Scientists studying the data discovered that there were notable differences in the brightness of different areas, which they concluded corresponded to the relative age of the lavas, with fresher (younger) lavas being brighter. But it is not possible to figure out actual, or absolute, ages of the rocks from this method. We would need physical rock samples to analyse in laboratories on Earth in order to do that because the regions are too small for accurate crater counting.

To attempt to shed light on the age of Venus' lavas – to see if any have flowed more recently – another group of scientists decided they would simulate the surface conditions at Venus. To do this they set up a chamber in a laboratory on Earth with conditions that were as similar as possible to those on Venus. Their aim was to see how different minerals that are normally found within Venusian lava flows react to its alien (to us, anyway) environment and, therefore, how they might change over time. These experiments showed that olivine, which is an abundant mineral in basalt rock, reacted much more rapidly to Venus' surface conditions than it would on Earth, where it is also very abundant. In these experiments, olivine was found to become coated in iron oxide minerals within just a few weeks. Fortunately, these iron oxides can be detected from space, and so, based on this work, when they are detected in lava flows it suggests to scientists that the olivine in the basalt has been exposed at the surface for at least a few weeks. Unfortunately, it's not possible to be any more precise than that. However, the most exciting part is that not all the basalts observed on Venus contain olivine that has altered to iron oxides on the surface, which suggests that some lava flows are literally just several weeks old. We can't be certain, however, that the experimental data and findings from this study accurately reflect the situation on Venus, because it is not possible to perfectly replicate all of the conditions in a laboratory on Earth. But if they do, then we can conclude that Venus is active today.

If that is not convincing enough for you then, thankfully, there is further evidence to support Venus currently being active. Since as far back as the 1970s and 1980s when the Pioneer Venus Orbiter was studying Venus, up to very recently when the Venus Express Orbiter was doing the same, there have been intriguing measurements of spikes in the abundance of sulphur dioxide within Venus' atmosphere. On Earth, sulphur dioxide is well known to be a major volcanic gas. A volcanic origin is the most likely explanation for the sulphur dioxide in Venus' atmosphere too. However, the fact that spikes in the abundance of sulphur dioxide were detected on Venus is key, because it suggests that the gas was originating from within the planet to account for these short-lived increased abundances. With no evidence of active volcanoes, scientists couldn't be sure if this was true. However, the evidence for active volcanism on Venus seems to be gathering pace and, if the laboratory experiments are correct, the episodic spikes could be related to the very recent eruption of lava flows on the planet.

If there are lavas being erupted at the present day on Venus, the likelihood is that they are relatively small in size, particularly in comparison to those that appeared around 500 million years ago. But it might be that Venus could experience another major period of volcanic activity in future years. The reason scientists think this might be a possibility is because of Venus' stagnant lithospheric lid, which continues to act as a blanket, keeping the planet's insides warm. Spacecraft have measured the heat flow from Venus and it is estimated to be half that produced by radioactive heat-producing elements within the planet. This means that the heat Venus' rocks are producing on the inside is not necessarily getting out, further suggesting that Venus may still be heating up. It's hard to know when, but if this situation continues, and Venus' insides get warmer and warmer over time, our planetary neighbour could see another major overturn event in the future, with its surface experiencing a huge influx of flood plain lavas that could, once again, cover over the present-day geology.

Whereas plate tectonics can account for about 90 per cent of Earth's volcanoes, on Venus there is obviously another explanation for its highly volcanic surface. Venus' volcanoes are almost certainly fuelled by mantle plumes, much like those in Hawaii and other volcanoes related to hotspot regions on Earth where columns of hot rock rise from the depths of the planet. In particular, there are three so-called 'rises' on Venus, called Dione, Imdr and Themis Regiones, that have been identified as sites of mantle plumes upwelling, in particular because of their raised topography. Scientists who have produced computer models of the inside of Venus suggest the planet might have nine of these major mantle plumes that rise from its deep interior all the way to the surface. This is similar to the number of mantle plumes thought to exist within our own planet's mantle.

Despite the fact that Earth and Venus both have mantle plumes, the lavas that have flowed from Venus' volcanoes haven't behaved like our own. Spacecraft images have revealed what look like the remnants of rivers that ran across Venus' surface. These features can extend for hundreds of kilometres. Were it not for the fact that we know Venus' surface is sitting at temperatures high enough to melt lead, scientists might have thought these channels were carved by rivers of water. As it is, they've figured out the features that look like sinuating rivers are instead exceptionally long solidified lava flows. These flows extend way beyond even the largest and longest to have flowed in Earth history, or at least those that have been preserved in the geological record. The explanation for Venus' lava flows having travelled such great distances will come as no surprise: it's thanks to its exceptionally hot surface temperature. The lava simply doesn't cool as quickly as it would on Earth once erupted, so it remains molten for much longer and is capable of flowing further.

To truly find out if Venus is still active at the present day requires further space missions. There is every chance that Venus' volcanoes still flow, but they might do so infrequently, so catching them in action would be down to luck. The next best thing is to place seismometers on the surface, much like the NASA InSight mission has done on Mars. However, Venus' surface conditions are not just inhospitable to life; scientific equipment will also have a hard time surviving the high temperatures. We saw that the Russian Venera landers lasted between just a few hours and a few days on Venus in the 1970s before being crushed to death.

The challenges in studying even the closest of planets to Earth are very apparent here. Venus may be a planet much like our own, being born of the same materials, in roughly the same part of the Solar System, but it looks and behaves very differently.

## The red one: Mars

Olympus Mons is one of the most famous of the Solar System's volcanoes because of its enormous size, measuring over 800 kilometres (500 miles) across and reaching a height of 21 kilometres (13 miles) – about two and a half times higher than Mount Everest. But as we've seen, Mars' biggest volcano is formed from the very same material that produces many of Earth's volcanoes and most of the others around the Solar System: basaltic lavas. As far back as 1972, Mariner 9 revealed that a large portion of Mars is covered in volcanic rocks that make up extensive lava flows and plains, and the largest volcanoes in the Solar System. These rocks give Mars its red colour, or rather, it comes from the dust that is produced as the rocks are weathered over time. Mars' basaltic dust contains more iron than the equivalent basalt on Earth, and it is this that turns to rust because of Mars' low atmospheric pressure: radiation from space can reach the surface and cause oxidation, thus making a red planet with dusty, red skies. But if you were to scratch the surface you would find either hard black basaltic rocks underneath, or – as NASA's Curiosity mission has shown – sedimentary rocks ranging from grey to red in colour, made of all the ingredients from the basaltic precursors plus some additional minerals.

Mars' surface geology can be split in two, broadly divided into the northern and southern hemispheres. The southern hemisphere, called the southern highlands or uplands, appears to be much older based on the heavy cratering seen there. The crust in this hemisphere is much thicker and sits topographically higher than the northern half of the planet. The northern hemisphere, also known as the plains or lowlands, is an enormous topographic depression sitting at a much lower elevation. This region is also much less cratered, indicating that it is younger in age and the exposed surface is not volcanic. Instead, the large basins that form the lowlands are filled with sediments. While these hide the underlying geology, in places, where it is occasionally exposed by impact craters, it is thought to be broadly basaltic in composition.

Just like the Moon, volcanism on Mars is very old, possibly dating back to at least four billion years ago, although the activity at the large shield volcanoes we still see today is more recent, at around one to two billion years old. Olympus Mons, the highest and most famous of the Martian volcanoes, is certainly not alone. It sits around 1,200 kilometres (745 miles) away from the Tharsis region, a huge bulge in Mars' lithosphere that occupies around 25 per cent of the planet's surface, between the southern and northern hemispheres. The Tharsis region houses three enormous volcanoes: Ascraeus Mons, Pavonis Mons and Arsia Mons, collectively known as the Tharsis Montes. While they are all smaller than Olympus Mons, the Tharsis Montes still range in height from 14 to 18 kilometres (8.5 to 11 miles). These volcanoes are aligned in a north-east to south-west direction, the reason for which is unknown, but it may be related to a structural feature within the Martian crust. The Tharsis region is thought to be very old, dating back as far as 3.7 billion years, but it also contains much younger flows. In fact, so much lava has been erupted in this region (hence the bulge) that it has placed immense stress on the lithosphere, resulting in extensional fractures and the creation of rift valleys. One region where this is evident is within the Valles Marineris, an extensive system of canyons running for 4,000 kilometres (2,500 miles) around the equator of the planet, 200 kilometres (120 miles) wide and 7 kilometres (4.3 miles) deep.

It's not just effusive, voluminous, runny basaltic lava flows that formed the Tharsis bulge; there are likely to have been countless generations of ash too. Yes, that's right, there's ash on Mars. It means the planet must have experienced explosive volcanism in addition to runny basaltic lavas, which scientists surmise occurred early in its history. However, evidence for this in the rock record is hard to come by because it is thought that much of this material has been covered over by more recent, albeit still ancient, flows.

As we've seen countless times, to produce explosive eruptions we need a viscous magma that can hold gas within it. Therefore, knowing whether Mars has experienced explosive eruptions informs scientists about the chemical history of the planet too, providing information about the volatiles, such as water, held within the interior. Interestingly though, Mars' low atmospheric pressure will also have influenced the type of eruptions experienced. A thin planetary atmosphere allows the gas within a magma to expand much more violently when it reaches the near-vacuum at the surface. Such rapid gas expansion within a magma will blow apart the molten rock, transforming it into finer particles such as ash, allowing it to spread further, and more quickly. There is some evidence for pyroclastic flows on Mars, but more work is needed to investigate and understand these features in detail. Even so, it is estimated that a pyroclastic flow could travel three times as far on Mars as it would on Earth because of the difference in surface conditions.

In particular, it is the features named paterae that have led scientists to think explosive pyroclastic eruptions were a common feature of Martian geology in the past.[*] Nevertheless, some detective work is needed to figure out what has happened, as they can only be studied from orbit. The paterae are not geographically close to the Tharsis region; instead they sit in the cratered uplands of the southern hemisphere. These volcanoes seem to have no Earth equivalent. They are relatively flat, being around 200 to 300 metres (660 to 1,000 feet) high and 1 to 2 kilometres (0.6 to 1.2 miles) across, with a central caldera surrounded by sets of radial furrows. Scientists think these furrows probably represent channels eroded into the volcanic deposits and that they are, therefore, probably made of thin piles of fine-grained and easily eroded pyroclastic material – ash – that has piled up close to the vent.

From the available measurements taken from orbit, it has been determined that the paterae are formed from basalt. But this basalt doesn't appear to be like other basalt we've met that flows to form flat sea-like plains. Instead, it seems to be stickier. The problem is that scientists don't yet have a clear understanding of why the basaltic lava in these locations is so sticky. One option is that the basalt interacted with water to produce a phreatomagmatic eruption, similar to those we find in volcanoes in very cold water on Earth, which are known to be very explosive indeed. In fact, the formation of the island of Surtsey, as described in Chapter 4, involved phreatomagmatic eruptions when the magma met the seawater, producing explosive volcanic activity above the sea surface.

---

[*] Paterae are shaped like shallow bowls. They are usually volcanic craters, but are sometimes formed by impacts.

Of course, for such an event to have occurred on Mars, it would need liquid water. While there is apparently no liquid water flowing on Mars' surface today (that we know of), water is known to have flowed on Mars in the past, when its atmosphere was thicker and able to support it on the surface. The explosion caused by a hot magma meeting cold water could definitely account for pyroclastic deposits spreading as far as 600 kilometres (370 miles), which is what has been observed in some locations on Mars. In addition, even at these large distances from volcanic vents, deposits have been estimated to be 150 metres (almost 500 feet) thick. This means they are highly unlikely to be wind-blown deposits, which would be thinner, and are much more likely to have been emplaced by pyroclastic flows. Without sufficient air pressure and resulting wind that can carry lots of material, it would be hard to move large volumes of ash over such long distances. While Mars might seem like a cold, dry and quiet desert today, it has certainly seen its fair share of action in the past.

Mars also had fire fountains, which as we've seen are jet-like sprays of lava. The highest fire fountain recorded on Earth occurred on the Japanese island of Oshima in 1968, producing a spray of lava reaching up to 1.6 kilometres (1 mile) into the air. Yet, on Mars, fire fountains could have been twice as high, thanks to its lower gravity, which is due to its smaller size. Another effect of lower gravity is that the magmatic plumbing system feeding magma under the surface is expected to be larger. Wider magmatic feeding pipes would have meant higher eruption rates compared with those on Earth because the pipes allow the transport of larger amounts of magma in a shorter time. But the Martian lava flows produced would also be expected to be much longer than those on Earth because the higher eruption rates would mean flows could be thicker, so they wouldn't have cooled as quickly and would travel further. Even a small change in surface conditions can have a big effect on the deposits we see.

Over the years, as the resolution of imaging by spacecraft in orbit around Mars has improved, it has allowed scientists to identify increasingly smaller features on the surface. Some of these features are pyroclastic cones, which are the most common type of volcano on Earth. They form from the piling-up of erupted materials close to a volcanic vent and have steep sides of around 30 degrees. The eruption that produced them will have been explosive, characterised by viscous, sticky lava that can't flow very fast. The problem for scientists is that although pyroclastic cones have been identified on Mars, they don't seem to be as common as expected based on the other explosive activity that has been identified.

Moving on to more recent times, Mars appears to host some younger and smaller lava flows. Some of these are found on Olympus Mons and are just 20 to 200 million years old, potentially representing the last gasp of Martian volcanism. There is currently no evidence for any active volcanism taking place at the present day and it is thought to be unlikely that Mars will have any significant volcanism in the future.

One of the reasons Mars may have seen its last days of volcanic activity is because of its lower mass, which is about 10 times less than that of Earth. Mars' current internal heat production is, therefore, much lower because it originally started out with less radioactive heat and less primordial heat from its formation. Mathematical thermal evolution models predict that volcanism should have stopped at least a few hundred million years ago. However, as we've seen above, these models are contradicted by several studies that seem to indicate volcanic activity has occurred in the last tens of millions of years. Is it true that Mars is not yet volcanically extinct? Only time will tell...

This is partly why NASA launched the InSight mission in 2018, which landed on the red planet later the same year. One of its aims is to search for tectonic activity, to see if the insides and surface of the planet are still moving. And it turns out that while Mars may be very cold and desert-like on the surface, its interior is far from dead. NASA's InSight lander showed us that Mars is still tectonically and seismically active, measuring hundreds of marsquakes over the course of a year. It's not quite as active as Earth, but it is much more active than the Moon. In fact, Mars' quake activity has been likened to that experienced by portions of our own planet that are located far away from tectonic plate boundaries, in regions classed as 'intraplate'. Of course, Mars has no plates – it is a 'stagnant lid' planet – so where do all these quakes come from? The quakes InSight has measured on Mars almost certainly come from the continued and gradual cooling of the planet, which causes contraction in the crust as it accommodates the shrinking.*
Mars' quakes are all small, only reaching magnitude 4 on the Richter scale. But of these quakes, some of the largest are located in a region called the Cerberus Fossae fracture system. Crater counting has revealed this to be one of Mars' youngest regions, with lavas possibly having flowed within the last 10 million years. In this region, marsquakes are still thought to indicate contraction, but here they might be related to the existence of a body of magma beneath the surface, one that is still cooling. While it is unlikely that this body of magma could be warm enough

---

* Some of the marsquakes InSight measured were caused by meteorite impacts onto the surface of Mars.

to produce an eruption in the future, scientists cannot be 100 per cent certain that it couldn't. There is still much for us to learn about this planet that has never been visited by humans.

As we've seen, scientists are quite certain that, like all the other planets except Earth, Mars doesn't have plate tectonics and the quakes that have been measured can happen without the movement of plates. This means that Mars' volcanoes are fuelled by mantle plumes and not plate boundaries. As this is the case, there are no volcanic chains on Mars like the ones we so commonly see on Earth. As we saw in Chapter 4 and earlier in this chapter, the Tharsis volcanoes do appear to be aligned on the surface, but each is thought to have been fuelled by its own mantle plume, meaning they are not otherwise related to one another. The large size of these volcanoes actually helped scientists to conclude that they sit atop their hotspots for ever and are not moved away by lithospheric motion, allowing them to grow very large over time as lava flows pile up on top of each other.

Nevertheless, there is some intriguing evidence to suggest that the red planet saw plate tectonics in its past. In Mars' ancient highlands, scientists have reported observing features within the rocky landscape that show parallel bands of alternating polarity. These magnetic stripes are somewhat similar looking to those that form around seafloor spreading centres on Earth, like the one in the middle of the Atlantic Ocean. On Earth, we've seen that these stripes form on each side of a mid-ocean ridge as new lava is erupted and the iron minerals within it line up with the planet's magnetic field. When the magnetic field reverses, which last happened around 780,000 years ago on Earth, the iron in the minerals within the lava being erupted then aligns to the new magnetic field, which is different by around 180 degrees. The reason we have this massive dynamo within our planet is because of our giant liquid metallic core surrounding our solid metallic inner core. But today Mars has no inner dynamo to create a global magnetic field, and we know this because we haven't detected one. The magnetic stripes, however, could suggest this hasn't always been the case.

While this evidence is exciting, it is unclear whether Mars' magnetic stripes really do reveal ancient plate motions. Firstly, the stripes do not form obvious symmetrical patterns like those we observe on Earth. They also don't appear to be offset by fractures along the length of the supposed ancient ridge, which are necessary to allow for the curve of a planet's surface as the ridge extends around it. But an explanation is still sought, even if the stripes aren't perfectly aligned in the way we expect for seafloor spreading. What we do know is that these features are very old indeed. Crater counting in this region reveals they date back to at least the Late Heavy Bombardment phase of Solar System history, around four billion years ago. If plate tectonics did happen on Mars, then the evidence here suggests that it must have ended by 3.8 billion years ago, meaning it can only have occurred over a very short time interval, if at all. Plate tectonics on Mars is unlikely to have developed to the point where subduction took place to produce active volcanoes. But if Mars did have some form of very primitive plate tectonics, then it would make it the only other rocky planet to have done so. The implications of this are wide-reaching, because it is thanks to the confluence of several factors on our planet working together that it is hospitable to life. Along with water and a protective magnetic field, plate tectonics is also probably necessary, so there is an intriguing possibility that our closest neighbour could have hosted life in the past.

## Mars on Earth

We are fortunate that, unlike Venus and Mercury, Mars provides us with samples as meteorites arriving on Earth, so we can study its geology the same way we do for the Earth and the Moon. These samples can provide valuable insights into the rocks that exist on the surface which we would otherwise only be able to study remotely using robotic landers and orbiting spacecraft. When meteorites presumed to have originated on Mars were first studied, they were divided into three rock types known collectively as the SNCs, which stands for shergottite–nakhlite–chassignite. At the time, it was hotly debated whether these really came from Mars. First, they were young, some as young as 180 million years. As we've seen in Chapter 9, this ruled out the Moon as the meteorite source. They had to have come from a planet that kept its heat for much longer. But which planet? One way to investigate the origins of these strange rocks was by using noble gases: those elements that are well known because they don't react with anything. Noble gases had been measured on Mars by the Viking landers in the 1970s, and were found to be special. One specific isotope, xenon-129, is much more abundant in Martian atmospheric xenon than in xenon from Earth, or any other meteorite that had been found and measured to date. This noble gas 'fingerprint' of the Martian atmosphere was also found in glasses in the shergottite meteorites. While the xenon in the meteorites fitted the Martian values measured by Viking, the open question was whether Mars is the only place with this strange and different xenon. Continued measurements of samples, including those made by the Mars Exploration Rovers from 2001 onwards on the planet's surface, gradually filled in the picture enough to confirm that the rocks we thought are from Mars really are. Despite humans not having been able to visit and bring back samples from the red planet, the cosmos has delivered them for free in the form of over 100 different meteorites, most of them magmatic rocks. The Martian meteorites are chemically and texturally similar to Earth's basalts, but they differ in having more iron and volatile elements, and lower nickel. The Martian meteorites have been radiometrically dated and those that we can put into the category of 'lava flows' (nakhlites and shergottites) are between 1.4 billion and 180 million years old, indicating the time when they crystallised as a solid rock. Because of these relatively young ages, they are considered to have come from the lightly cratered volcanic regions such as the Tharsis plateau.

As we all know, not everyone fits into the family picture, and the Martian meteorites are no exception. There is one Martian meteorite, nicknamed Black Beauty, official name North West Africa (NWA) 7034, that isn't a magmatic rock at all. Originally found in the Sahara Desert in 2011, it was soon obvious that this rock was very special, not only having come from Mars, but looking nothing like those that had been collected before. Subsequently, Black Beauty developed a sale price of \$10,000 (£7,500) per gram. But why is this rock so special? Black Beauty represents a part of Mars from which we currently have no other meteorite samples. Of course, we haven't sent any humans to sample Martian rocks yet, and we haven't had any robotic sample return missions either, so our samples from Mars are biased towards whatever space has sent us. Black Beauty is a breccia – a rock consisting of lots of different sharp fragments of other rocks – and has been found to contain some clasts* that are 4.4 billion years old. These ancient samples of rock are thought to be pieces of Mars' early crust. Intriguingly, this means that Mars' first crust formed much earlier than Earth's, just 20 million years after the formation of the Solar System. We still have a lot to learn about Mars, and more samples like Black Beauty will help us, but it has already shown us that Mars' cooling history has differed from our own planet's and might explain more about the volcanism that has taken place over the course of the four billion years since. To learn more about Mars we will have to send robots to return samples and, even better, humans, with their ability to explore the surface, scouring it to collect rocks in the same way we have done on the Earth and the Moon.

---

* A clast is a fragment of a rock or mineral ranging in size from a fraction of a millimetre to as big as a large building.

# CHAPTER ELEVEN

# Ice Worlds

Before 1610, we knew nothing about the worlds – or rather moons – existing around other planets. The Earth was thought to be at the centre of the Universe and our Moon unique. When the first four moons of Jupiter were discovered by Galileo Galilei, nothing was known about them, but the one thing we thought we knew was that they would be cold and dead. After all, these were tiny worlds, a huge distance from the Sun. It was completely unexpected when we discovered in more recent years that not only are there hundreds more moons out there (Jupiter alone has 79), but that some of them are active. Sure, they might be cold, but these worlds are alive!

We met Io in Chapter 9 and saw how it is a lively world of 'fire', erupting hot silicate rocks but with ice covering its surface. Yet Io is an oddball, being the only active silicate, rocky object in the outer Solar System. The other worlds that join Io in the outer Solar System are quite different, having crusts made of ice, but that doesn't mean they're not active. Many of these worlds comprise rocky metallic cores and can produce heat in their interiors, which means they are able to create activity at their surfaces. While these places possess volcanoes made of ice, which makes them seem wildly different to the worlds with volcanoes made of fire, these ice worlds are not as different as you might think.

## The first: Triton

The ice worlds of the Solar System were long thought to be dead, literally frozen in time and unchanged from their formation, except for the scars produced by impacts of comets and asteroids over the years. This was until we got closer to them. As we've seen, it was the Voyager spacecraft that first revealed the unexpected activity taking place at some of these mysterious frozen worlds.

The first world on which we saw unexpected activity was Neptune's largest moon, Triton, the seventh largest moon in the Solar System. The surface of Triton is very rugged in places, but despite this feature normally indicating an old surface that has been bombarded by space rocks, Triton is sparsely cratered. Instead, its ruggedness looks more like that seen on Earth. Its lack of craters, and the addition of some very smooth plains in its southern hemisphere, indicate a world that is active.

However, the surface of Triton sits at -235°C (-391°F) so it should be frozen solid and dead. In 1989, however, Voyager returned images of icy plumes erupting nitrogen and methane from Triton's southern hemisphere surface. These plumes weren't insignificant in size either: they had diameters up to 1.5 kilometres (0.9 miles); they rose to altitudes of 8 kilometres (5 miles); and they created deposits that were strewn out over Triton's surface for around 150 kilometres (93 miles). Triton was the first example of cryovolcanism in the Solar System, revealing an active frozen world erupting its own brand of magma – magma mainly made from ice – into space.

We have seen that there are many different types of silicate volcanism, from explosive to runny, and cryovolcanoes can vary in composition just as much. While cryomagmas are of course much cooler than silicate ones, they still have the power to sculpt landscapes in a similar way. A cryomagma can be made of water, or other volatiles such as ammonia and methane. Depending on the exact mix, and the environment into which they are erupted, they can take on very different properties, being frothy, dense and/or runny.

Triton's surface is covered with annealed frozen nitrogen, with a nitrogen and water-ice crust and an icy mantle, yet deep within it is made of dense silicate rock and metal, making it a differentiated object much like Earth and the other terrestrial planets. Based on density measurements, Triton's core constitutes a substantial part of the moon: around two-thirds of its mass. Its rocky insides are thought to contain a substantial number of radioactive elements, enough to generate the heat that maintains a subsurface liquid ocean below the icy crust. But those radioactive elements can't account for all of Triton's heat because they are not thought to be enough on their own to have melted and kept an ocean in a liquid state for a long time. There is another source of heat within Triton, but to understand what that is involves a dive into its history, all the way back to its early beginnings, when it wasn't a moon at all.

Scientists think that Triton may have originally started out life in the Kuiper Belt, which is home to millions upon millions of icy objects outside the orbit of Neptune. The Kuiper Belt is the origin of many of the short-period comets that enter the inner Solar System and transit close to the Sun every now and again, but it is also home to the dwarf planet Pluto. While Triton is a little larger than Pluto, it is of a similar composition, and the two may be related way back, billions of years ago, originating from the same Solar System neighbourhood. Triton and Pluto also have something else in common: Triton may have once had a friend with it, another object that joined it in what is known as a binary system, just like Pluto and its moon Charon. However, early in Solar System history, as Triton went about its orbit around the Sun, it may have got dangerously close to Neptune, such that it was pulled in by the giant planet's gravity. Triton was then orbiting around Neptune and not the Sun, and in the chaos, it lost its little partner.

This theory of Triton being captured by Neptune is used to explain why it has a strange orbit around its planet. Triton's orbit is classed as retrograde because it goes in a different direction from the orbit of Neptune. This characteristic suggests the two objects didn't form together and instead that Triton was a foreign object that arrived later. But you might still be wondering how this origin relates to Triton's internal heat. When Triton was first taken hostage by Neptune's gravity it is thought to have had a very elliptical orbit around its new host planet. The result is that, as Triton travelled around Neptune, it was sometimes closer and sometimes further away. The effect of this to-ing and fro-ing is tidal heating. As Neptune's gravity pulled and squashed Triton during its orbit, its insides were heated by friction. Over time, Triton's orbit has settled down into a more circular one, causing less tidal heating, but the effects of the earlier tidal forces were preserved for billions of years and, to this day, the current effects of tidal heating continue to keep it warm.

The thing about heating Triton to produce cryovolcanic activity is that, because of its composition, not much heat is needed to make something happen. A mix of ammonia and water can melt at a very low temperature and nitrogen compounds can change from solid to liquid and gas at temperatures less than -300°C (-508°F). In fact, a temperature rise of less than 10°C (18°F) might be enough to make a water–ammonia mixture flow like a very cold, sludgy lava resurfacing the moon to form smooth plains. In addition, snowflakes of nitrogen can form from Triton's plumes, as the material is carried downwind to create nitrogen snowfall back on to the surface.

The other feature commonly seen on Triton's rugged surface is what is described as 'cantaloupe' terrain, as it resembles the surface of a cantaloupe melon. The features that form this interesting pattern on Triton are made up of irregular mounds a few hundred metres high and a few kilometres across, interspersed with large depressions tens of kilometres in size. The terrain is thought to be the result of rising blobs of ice, otherwise known as diapirs, from Triton's interior. These diapirs, being less dense than the overlying ice crust, work their way up (a little like a rocky magma rising through the Earth's crust) to reach the surface, producing the cantaloupe pattern. This type of process is only possible if the surface is being heated from below, providing further evidence that Triton is still active within.

One of the exciting prospects for worlds like Triton is that, despite its frigid surface, the fact that it contains a liquid ocean gives it the potential to host life. While we think liquid water is one of the most important ingredients for life, it might be that any liquid is capable of supporting life; we just don't know for sure yet what is possible. So, while it might be a stretch of the imagination to think of life lurking under the thick shell of an icy crust on a small moon billions of kilometres from the Sun, it doesn't necessarily have to be anything like us. In fact, there is no oxygen in Triton's atmosphere, so there is no chance of something human-like anyway.

However, as we've started to explore the base of our own planet's oceans, we've seen that not all life needs the same ingredients we do. All life as we know it needs water, carbon, a source of energy and a few other elements (for example, phosphorous and sulphur are useful as they are key to building the macromolecules that carry genetic information). But we also know that, beyond the basic elements, things get very diverse. While plants on the surface get their energy from oxygen and sunlight, as we've seen, some of the life existing in the depths of Earth's oceans doesn't require oxygen or sunlight, simply feeding off the geochemical energy the planet gives out in the dark, seemingly inhospitable black smokers on mid-ocean ridges. It'll be a while before we find out if Triton hosts life since, at the time of writing, while a mission to visit Triton is in the evaluation stage, nothing has yet been decided. It will be a while before we can visit the outer planets, but they are certainly something we can continue to speculate about.

## The habitable one: Enceladus

Enceladus, the sixth largest moon of Saturn, and about 500 kilometres (300 miles) across, turns out to be the smallest cryovolcanically active object in the Solar System. Around 2005, NASA's Cassini spacecraft photographed plumes on the south pole of Enceladus shooting water vapour into space. Enceladus is one of the most reflective bodies in the Solar System thanks to its fresh, smooth, icy surface made mostly of water. While some its regions are marked by rugged mountains, others are heavily cratered, and others again are covered in crater-free, smooth plains. As far back as the 1980s, Voyager flew by to reveal Enceladus' relatively young-looking, yet complex, surface cut by large fractures, leading scientists to postulate that it might be geologically active. But its daytime temperatures reach just -198°C (-324.4°F), and its small size meant that scientists were very surprised when they found it spewing material into space.

There is something very helpful about plumes like the ones on Enceladus because they move material from the inside out into space. This gives us a chance to sample the interior of a new world without having to go to the trouble and expense of sending a lander down to the surface. In 2015, Cassini 'simply' flew through Enceladus' plume, effectively tasting it to tell us what was there. What it found was a whole host of molecules including water, hydrogen, grains of silica, hydrocarbons and even sodium chloride crystals, that is, salt. Subsequently, hundreds more plumes have been identified on Enceladus in its south pole region and much more is understood about their formation. Some of the plumes emanate as jets, and others from fissures similar to those found commonly at Hawaiian volcanoes when hot lava is released in an apparent curtain, except on Enceladus, we are obviously looking at freezing cold materials instead. The regions on the surface that they blast from are known as 'tiger stripes'. The stripes are around 130 kilometres (80 miles) long, 2 to 4 kilometres (1.25 to 2.5 miles) wide and are flanked on each side by tall ridges around 100 metres (328 feet) high. They are each separated by about 35 kilometres (21 miles).

Much like Io in the Jovian system, Enceladus gains its heat from tidal forces because it is trapped in an orbital resonance with another moon, Dione, which leads to a slightly elliptical orbit around its planet. Consequently, it is sometimes closer and sometimes further away from Saturn, whose gravity pulls and tugs at its insides to produce heat and activity. The main difference is that this heating doesn't reach the same high temperatures as Io's insides, but that doesn't mean activity can't happen. Another effect of these tidal forces is that, as Enceladus is pulled around, its fractures and tiger stripes are opened and closed, which allows jets and curtains of plume material to shoot out into space. Instruments on the Cassini spacecraft measured temperature differences across the tiger stripes to find that the fissures are warmer than the expected values for the region: -80°C (-112°F) as opposed to -230°C (-382°F). Although these temperatures are still on the cold end of the scale, it doesn't matter, because the materials that make up Enceladus can change phase, even at these low temperatures, to create activity. The heat coming from Enceladus' interior provides a continuous release of energy. Here on Earth, this energy would be enough to power a city of around six million people.

Much of the material that Enceladus produces from its plumes falls back onto the moon's surface, including Enceladus' own style of 'flood basalt' volcanism, where huge volumes of icy flows are thought to cover parts of the smooth plains. But not all of Enceladus' cryovolcanics end up staying on the surface. In fact, it is thanks to Enceladus' cryovolcanoes that Saturn hosts its E-ring, the widest and outermost of all the rings, within which Enceladus orbits. The plume materials that are energetically pushed into space at speeds of 400 metres (1,312 miles) per second, overcoming Enceladus' gravity, end up forming the E-ring. It consists of tiny particles of water ice, silicates, carbon dioxide and ammonia. The activity on Enceladus continuously replenishes the E-ring with fresh materials.

At Triton we discussed the potential for life within a subsurface liquid ocean. Evidence for the same situation can be found within Enceladus, and the potential for life is possibly even higher. Heat produced within Enceladus, just like on Triton, is thought to maintain a liquid ocean beneath its icy crust. Scientists have been able to hypothesise this thanks to the observation that Enceladus wobbles as it orbits Saturn. The wobble is purported to be because of a substantial liquid ocean beneath the surface, which is decoupled from the icy crust and the rocky core below. A subsurface liquid ocean certainly helps to explain where all the material comes from to fuel the plumes, which themselves are another line of evidence in support of this idea. Scientists know that this ocean must be salty because salt crystals have been measured in the plumes, and the other components of greatest interest found within the plumes, if we are talking about life, are the hydrocarbons, as they prove that carbon is part of the moon's chemistry.

Along with the salts, the silica dust that is erupted out in plumes can only form at high temperatures of at least 90°C (194°F). Scientists think they must form where water meets rock, at the base of the ocean: the water trickles down into the porous, rocky core, where it becomes heated and initiates reactions. The silica grains themselves are thought to show specifically that hydrothermal activity is taking place. The water heated by warm rock is more buoyant, causing it to rise back up into the ocean above, where it precipitates out the minerals it has collected from the rocky core below. This is the same process as that occurring at the base of Earth's oceans along mid-ocean ridges. On Enceladus, the process continues, as the hot plumes of material rise from the deep throughout the colder ocean towards the icy shell at the surface, and they erupt when the ice is pulled apart. The heat energy emanating from just one seafloor hotspot is thought to be equivalent to the annual geothermal power consumed in Iceland.

The key thing about the rock–water–ice interactions within Enceladus' interior, though, is that they transfer energy and produce materials that might support life. On Earth, our deep-sea hydrothermal vents support ecosystems that are powered by chemicals and not sunlight. Is the same process happening within Enceladus? There is every possibility that it is. Molecular hydrogen is found in high abundance within Enceladus, along with carbon dioxide, which together produce methane, a reaction that is one of the most primitive metabolic pathways used by microbes. Such microbes are known as extremophiles because they can survive in conditions that we, as humans, consider extreme. But even if we thought that an icy moon orbiting a giant planet in the outer Solar System could have life, then we wouldn't expect it to look like us anyway; so alien life that doesn't even require sunlight might not be a surprise. Either way, the conditions and processes working together within Enceladus make a strong case for it being a habitable world. We just need more data before we know for sure.

### The Earth-like one: Titan

Titan, not to be confused with Neptune's moon Triton, is a moon of Saturn. It was the first Saturnian moon to be discovered, back in 1655, and is also the biggest. Titan is larger than the planet Mercury and about 1.5 times larger than our own Moon. Its size is an important feature, because it means it can hold on to an atmosphere, one that is about 60 per cent denser than Earth's and filled with nitrogen just like ours, but containing very little oxygen. When scientists first figured out that Titan had a thick atmosphere, it was an intriguing feature because it opened up the possibility for liquid water being stable on its surface. If Titan hosted liquid water reservoirs on its surface, then it would be the only place in the Solar System other than Earth to do so. Yet, if there was liquid on the surface of Titan, it couldn't be water because its surface sits at around -180°C (-292°F), far too cold for liquid water.

While Voyager 1 and 2 were the first spacecraft to get up close to Titan, they saw only the thick orange haze that makes up its impenetrable atmosphere, which was similar to the view of Venus from space. It wasn't until the Cassini spacecraft was launched in 1997 on its mission to the Saturn system that we really started to discover the true nature of this secretive world. This was because, thanks to the information supplied by Voyager, Cassini was appropriately equipped with a radar to peer through Titan's thick smog. Cassini performed more than 80 fly-bys of Titan, and it was during four of these that Cassini's Radar Mapper revealed Titan's surface to be geologically complex, not only showing geological processes that must be happening now or very recently − determined from a lack of cratering − but also showing evidence for cryovolcanism. Cassini imaged a region on Titan known as Sotra Patera: a depression, or pit, 1.7 kilometres (1 mile) deep and oval in shape, suggesting it is not an impact feature. The region also hosts Titan's highest mountain, known as Doom Mons, which is around 70 kilometres (43 miles) in diameter and nearly 1,500 metres (4,900 feet) high. Importantly, emanating like fingers from Doom Mons are what appear to be lava flows. Scientists think that this complex region of Titan provides the best direct physical evidence for cryovolcanism, even if we haven't seen it erupt in real time.

Cryovolcanoes on Titan would certainly help to explain a long-standing mystery about the moon's atmosphere. While it is mostly composed of nitrogen, there are substantial amounts of methane too. The interesting thing about the methane is that its abundance is not decreasing over time. Scientists know that methane high up in Titan's atmosphere should be broken down into simpler molecules by sunlight, on a timescale of around 10 million years, thus decreasing the amount of methane over time. Yet this isn't the case. For a long time, scientists suspected that Titan might have ice volcanoes that pumped out methane during eruptions to replenish that lost in the atmosphere. Luckily, the Cassini mission had another string to its bow that would help provide proof: the Huygens lander.

Cassini's buddy, the Huygens lander, hitched a ride from Earth to Titan, before being released by the orbiter to venture down to the surface and punch through Titan's thick atmosphere. Much like with Venus, to peer down to the surface of Titan requires either going there with a lander or using radar vision. The Cassini mission did both. But when scientists planned the mission, they had no idea what to expect when Huygens landed, as they'd never seen the surface. Was Titan covered in large oceans, or was it bone dry? The Huygens lander was designed appropriately, to be capable of landing on any surface.

The life of the Huygens lander was certainly shorter than that of the Cassini orbiter, whose reign over the Saturn system extended to 13 years. Huygens couldn't stay powered for very long on Titan's surface and the lander lasted just 72 minutes after touchdown, during which it transmitted a wealth of data back to Earth before its battery energy ran out. From the information gleaned during this short time, combined with measurements taken during its over two-hour descent to the surface, scientists were able to learn a great deal about Titan. As Huygens descended through Titan's atmosphere it measured an increase in the humidity and, as it reached the surface, an increase in the abundance of methane. Scientists concluded that this meant Titan was covered in liquid, in the form of seas and lakes, but that they must be made of hydrocarbons (molecules of hydrogen and carbon) such as ethane and methane because the moon is too cold for liquid water.

Huygens never saw flowing or pooling liquid on Titan's surface, but what it did see were the recent effects of flows that had carved canals, riverbeds and ephemeral lakes into the surface. In fact, Titan's surface, while made of ice, acts more like rock. There were all manner of fluvial features and even 'sand' dunes made of hydrocarbons that are thought to look like coffee grounds. Titan experiences 'weather' much like our own planet, but a methanological cycle as opposed to a hydrological one. Scientists even think that it sees flash floods, where huge amounts of methane build up in its cloudy skies before raining out onto the terrain to create the features that we can see. Without the volcanoes continuously releasing volatiles, such as ammonia and methane, from Titan's interior none of this would be possible.

Titan's cryovolcanics come from its liquid interior. Just as we've seen with the other active ice worlds we've visited, Titan is also thought to have a subsurface liquid ocean below its icy crust, and in many ways we should think of these ice worlds as ocean worlds instead. The ocean is thought to be mainly water, and it is also salty. At the base of the ocean scientists think there is a shell of water ice surrounding a rocky core with radioactive elements that provide heat. The ice is thought to be ICE-VII, which we met in Chapter 4, formed when water is compressed at very high pressures. And Titan is thought to be kept warm in the same way as many of the other satellites of the outer Solar System: via a mix of radioactive heating from deep inside its core and tidal flexing. In fact, as Titan goes about its orbit of Saturn, where it is slightly less than 1.19 million kilometres to almost 1.26 million kilometres (740,000 to 780,000 miles) away, scientists have determined that its surface elevation can rise and fall by 10 metres (32 feet), suggesting its insides are highly 'warpable'. This alone wouldn't be enough to maintain its liquid ocean, but heat coming from its rocky core would provide the rest of the heat needed. Scientists think that Titan's liquid ocean is located just 100 kilometres (62 miles) below its icy crust.

Titan's cryomagmas that erupt onto the icy surface are produced from the subsurface ocean, meaning it acts a little like Earth's mantle. However, at these frigid temperatures, volcano science is turned upside down. Magma made of water originating from an ocean of pure water would struggle to rise through the water-ice crust above because it would have negative buoyancy, so it would become stuck below the icy lid. For this reason, it might be that Titan's volcanoes spew out ammonia-water slurries that could also contain liquid methane. Such magmas are thought to be partly crystallised, and partially melted ice mixtures. The addition of ammonia to the mix not only adds buoyancy to the magma but it also acts as an antifreeze, enabling the flow of the cryomagma when it would have been impossible if it were just pure water under the same conditions. In fact, in experiments carried out on Earth, scientists showed that the addition of just 5 per cent of ammonia to water ice can reduce its viscosity by 100,000 times, turning solid ice into a slurry that flows like liquid rock.

If Titan's magmas are a mix of ammonia and water, then cryovolcanism on this alien world is thought to behave in much the same way as basaltic volcanism does on Earth. This means that features such as shield volcanoes and runny lava flows could form, and that the potential for explosive volcanism is limited. This is partly thanks to Titan's dense atmosphere, which would prohibit explosive activity. Yet, there are a lot of unknowns, and it is hard to predict how Titan's lavas would react to the surface environmental conditions because we still know relatively little about this seemingly Earth-like world.

Fortunately, we might get the chance to find out more about this moon soon. NASA is planning a mission to visit Titan in the 2030s, which sounds like a long time in the future, but this is the nature of planetary exploration timescales. At the time of writing, the Dragonfly mission will launch to Titan in 2026. It is a quadcopter that will soar around Titan's skies on multiple sorties to examine various sites in more detail, touching down in between. And from what we've learnt from the other ice worlds, you won't be surprised to learn that Dragonfly is going to search Titan for signs of life, as well as investigating the origins of life.

One of the reasons Titan is so interesting to scientists is because the conditions on Titan today are thought to be similar to those on the early Earth, when life was just getting started. In fact, despite the fact it sounds very alien, with volcanoes made of ammonia-water mixes, Titan might be the most Earth-like world in the Solar System. Titan has some key features that give rise to the potential for life. The chemical reactions that the Sun initiates high in Titan's atmosphere produce a range of complex hydrocarbons, or rather, organic matter, which could be the building blocks for life, or provide the chemical nutrients it needs. They are not of much use in the atmosphere, but fortunately, once the reactions between hydrogen, methane and nitrogen take place, the molecules that are made end up raining back down to the surface. The result is that Titan is covered in organic material. These organic molecules get everywhere, having made their way into Titan's subsurface water ocean too, which could, therefore, harbour a potential habitat for life. Because Titan is a warm, active world, the molecules, and possibly any life that has used them within Titan's interior, can potentially also make its way to the surface, again thanks to the cryovolcanoes. When it gets there, because of all the liquid on Titan's surface, this organic matter, or the organisms that might be using it, could even find another potential habitat on the surface of this moon.

Titan is an active world with a warm interior and it has geological processes that help to create weather and pools of liquid on the surface. It has an atmosphere that allows these processes to continue long term. This alien yet Earth-like world represents one of the most exciting places in the Solar System to search for life, so it is fortunate we are returning.

## The cracked one: Europa

Moving from the Saturnian system back towards Jupiter, we pass another very interesting ice world, Europa. In many ways it is similar to the active worlds we have already met: it has a global ocean underneath its shell of ice and it has features on the surface that indicate tectonic activity is taking place, just like Earth. Europa's cracked, but relatively crater-free, surface is young. It is estimated to be just 60–90 million years old on average, suggesting this world must have been active recently, geologically speaking. The Galileo mission imaged many landforms on Europa that looked like examples of cryovolcanic features, including apparent volcanic domes and what appeared to be lobate features that looked like the extrusion of magma onto the surface. While the origin of some of these features is still open to debate, they help to build the case for Europa being cryovolcanically active at the present day.

Europa's fractured surface is thought to be constantly on the move in a way that is very similar to Earth's lithospheric plates, except they are made of ice and not rock. Europa's icy plates appear to move around in relation to each other, often colliding, with some thought to even be subducted back into the liquid ocean below. Europa is possibly the

only other world in our Solar System where strong
evidence exists for plate tectonics, but where silicate rock
magmas are replaced by cryomagmas of water that are
erupted onto the surface instead of hot lavas. Despite
Europa being the smallest of the Galilean moons, a little
smaller than our own Moon, and very cold on the surface,
it is far from dead.

Europa's subsurface ocean was detected by the Galileo
spacecraft, but not directly. Instead, Galileo's results allowed
scientists to simply infer that it was there. As Europa orbits
Jupiter it passes through the planet's enormous magnetic
field. As it does so, it gently warps the field, but Galileo's
magnetometer was still able to detect the change. In order
for Europa to warp a magnetic field, its insides must contain
something that acts as an electrical conductor. The most
obvious material would be a salty liquid, leading to the idea
that Europa has a briny ocean below its icy crust. Such an
ocean is thought to be significant in size; it's estimated to
contain more than double the water present in all of Earth's
oceans.

Europa's huge subsurface water ocean is produced
because, like the other active ice worlds, it is warm deep
inside. Europa lives in the same neighbourhood as Io, and
the two take part in the same gravitational dance with
Jupiter. The pulling and tugging of Jupiter and its other
moons help to deform Europa's icy crust as its insides are
heated up by friction brought about by tidal forces, just like
Io. But as we've seen already, Europa also has a warm,
rocky core, the decay of radioactive elements within
providing more fuel for its warmth.

Europa's cryovolcanism is mainly water-based, but the
problem with pure water is that it freezes into ice as soon as
it meets the frigid surface, which doesn't get above -160°C
(-256°F). Sure enough there might be some other substances

within Europa's ocean that could act to lower the viscosity
of its water so that it could flow, but scientists are still
searching for clear evidence of effusive eruptions and large,
runny lava flows on the surface. Because of this, Europa's
cryovolcanic activity is thought to be broadly in plume–
type eruptions that occur episodically over time. Such
plume eruptions have been detected via information
gleaned from the Hubble Space Telescope. Over a 15-month
period, scientists studying the Hubble data saw three
occasions of supposed plumes erupting from Europa. These
plumes are thought to rise around 200 kilometres (125 miles)
above the surface before raining back down as cryoclastic
deposits once frozen. The observations indicate the plumes
are frequent, but are not very long-lived events. While we
don't currently have enough high-resolution maps of
Europa's surface to determine with confidence what is
happening with its cryovolcanics, this situation will
hopefully improve over the next few decades.

Whatever the case, the evidence to support the idea
that Europa is an active world is overwhelming, and its
warm, rocky core, coupled with its salty ocean, certainly
make a good case for the possibility of life. Just as we saw
with Enceladus, deep at the base of Europa's ocean, where
waters infiltrate the rock, there is the potential for the
formation of hydrothermal vents. The heated waters that
have interacted with the rock would be expected to rise
back out into the ocean loaded with chemical nutrients
that they have leached from the rock. This could be
the fuel needed to power the life processes of simple
organisms.

However, the problem with the Jupiter system is its
extreme radiation environment, as was highlighted by the
Galileo spacecraft that studied Io nearby. This makes for a
very inhospitable environment on the surface of Europa.

But if life were hiding deep inside Europa's oceans then it is thought that the icy shell of water ice above, which is probably only 15 to 25 kilometres (10 to 15 miles) thick, could afford it enough protection from the radiation.

Fortunately, we should get the chance to learn a lot more about this intriguing ice world in the near future. Both NASA's Europa Clipper mission and the European Space Agency's JUICE (Jupiter Icy Moons Explorer) mission are set to launch in the 2020s, with Clipper hopefully getting as close as 25 kilometres (16 miles) above the surface of Europa. Both spacecraft are equipped with instruments to image and measure the plumes. If they see any plumes, which it is hoped they will, we will get to find out a lot more about them. As we've seen with other active ice worlds, these plumes provide a window into the inner workings of Europa, allowing us to probe down into its depths without having to physically get down there, and in turn we will learn more about what is going on under the seemingly inhospitable icy crust.

## The furthest: Pluto

No discussion of ice worlds would be complete without a visit to the Kuiper Belt, home to the dwarf planet Pluto and its lesser-known moons, of which there are five: Styx, Nix, Kerberos, Hydra and Charon. Before the New Horizons mission flew past Pluto back in 2015 at some 84,000kph (52,000mph), very little was known about this distant and deeply mysterious world orbiting at an average distance of 5,906,380,000 kilometres (3,670,050,000 miles) from the Sun. If you could travel at the speed of light, it would take you about 4 hours 15 minutes to get to Neptune, the furthest planet from the Sun. Yet, most of the time – because Pluto's distance from the Sun varies a little during its orbit – to travel to Pluto would take you more than an extra hour. It took New Horizons nearly a decade to reach Pluto because we haven't yet figured out how to travel at the speed of light. And even then, the spacecraft was launched at just the right time, to take advantage of a fortunate alignment of the planets which gave New Horizons a gravitational boost that increased the craft's speed by 14,400kph (9,000mph) as it went past Jupiter. Without this, New Horizons' encounter with Pluto would have been delayed for another five to six years. Despite its rather fast and fleeting encounter with Pluto and its moons, New Horizons revealed a system of worlds that were much more interesting, and active, than had been forecast.

Pluto and its moons are primitive bodies, much like the other icy objects that share the Kuiper Belt region in which it orbits past our outermost planet. These objects are thought to have formed at the beginning of the Solar System, around 4.6 billion years ago, even before the planets, and thus represent the oldest objects to orbit our star. The Kuiper Belt is cold, and everything within it was long thought to be frozen in time since the formation of the Solar System. Sure enough, some Kuiper Belt objects sometimes find their way closer to the Sun as comets that get heated up, releasing their ancient inventory of frozen volatiles and dust as they transit through the inner Solar System. We've been able to send space missions up to visit some of these when they pass closer to Earth and learn about the materials they contain. But Pluto, the so-called King of the Kuiper Belt, thanks to its large size, will never be a comet, and so learning about this far-flung world is harder.

What New Horizons revealed at Pluto was a world frozen solid, but one that was not frozen in time. It was long known that Pluto was different from many of the other Kuiper Belt objects, being large enough to once be classed as a planet. Of course, now we know that while it is still the largest object in the Kuiper Belt, it is not the only large object out there. Other dwarf planets such as Sedna and Eris have since been discovered, meaning this very mysterious and far-flung part of the Solar System has become even more intriguing. But added to this was the fact that when New Horizons arrived at Pluto, it instantly found it to be covered in a range of surface features, with some regions on its surface being completely crater-free. It is hard to stress how very unexpected this was for an object this ancient and far from the Sun. It showed that something must be going on deep inside this world that was expected to be long dead. Pluto is not simply a cold relic from 4.6 billion years ago; it has clearly evolved over time, with evidence of recent activity having shaped its surface.

On Pluto, water ice forms its 'bedrock', but it is too cold to melt, making a 'crust' that behaves a bit like rock does on Earth. New Horizons revealed Pluto has ranges of mountains hundreds of kilometres long that were clearly formed through some sort of tectonic process. There are also two rather large mountains, Wright Mons and Picard Mons, located at the southern edge of the so-called Tombaugh Regio, named after Clyde William Tombaugh, who discovered Pluto in 1930, and also known as 'heart of Pluto' thanks to its shape. One of the most interesting things about these mountains is that they host what look like a central crater at their peaks. Put simply, on Earth, mountains with a hole in the top are usually volcanoes. Pluto's mountains tower 6,000 metres (19,000 feet) high, and span hundreds of kilometres across. They look suspiciously like the shield volcanoes seen so commonly on the rocky worlds of the inner Solar System. The similarity of Pluto's mountains to those made of basalt on Earth is striking and it has led scientists to suggest that Wright Mons and Picard Mons are examples of recently active cryovolcanoes. In fact, although we've talked about cryovolcanic activity across many icy worlds in the Solar System, none have provided such a clear example of a volcano edifice as Pluto's icy mountains. Even on Enceladus, where spacecraft have observed plumes emanating from fissures, flying through them to taste the chemicals they emit, we still haven't seen any landforms that can be as convincingly classed as cryovolcanoes.

Scientists think that one line of evidence pointing to Wright Mons and Picard Mons having been recently active is the sparse cratering on their surfaces. But another feature helped to support the theory, and it came in the form of a large red crack located close to Wright Mons, named Virgil Fossae. On closer inspection of the spectrograph data it was possible for scientists to disentangle the composition of the ice making up the red crack, concluding that the region was rich in ammonia. We already know that ammonia has the special quality of lowering the freezing point of water. Its presence on the surface of Virgil Fossae provided extra evidence that Wright Mons could erupt ice flows made up of an ammonia–water sludge mix that can flow like lava, oozing out like peanut butter. The presence of ammonia on the surface of Pluto is also a useful indicator that its cryovolcanic activity has occurred relatively recently, on geological timescales, because ammonia is a fragile molecule. Under these conditions, ammonia is not expected to survive on the surface for very long. As it sits on Pluto's surface in the weak sunlight that reaches this far out into the Solar System, the ammonia is broken down by radiation: ultraviolet radiation as well as galactic cosmic rays. For

Pluto to be covered in ammonia ice means that it is being replenished, probably thanks to cryovolcanoes pumping material out from its insides.

These potential cryovolcanoes are younger than the heavily cratered northern plains of Pluto, but not as young as the southern bright plains that make up Sputnik Planitia within the western lobe of the Tombaugh Regio. This region is crater-free and very geologically complex. Its surface is made mostly of nitrogen ice, but it has dunes. The dunes appear to be made of methane ice grains probably derived from nearby mountains, just like sand dunes on Earth. There is also evidence of large ice flows that are thought to be glaciers made of nitrogen. Despite the surface of Pluto sitting at -236°C (-392.8°F), making water ice as solid as rock, ices made of nitrogen, carbon monoxide or methane are 'soft' enough to flow. Even if these ice masses move very slowly, at no more than a few centimetres per year, they still have the ability to resurface entire regions in short geological timescales, just as we know from the movement of rocks on Earth. Below these flows of nitrogen and methane, Pluto is made up of water ice, yet it is coated with a thin layer of more volatile ices that are thought to have been emitted during volcanic eruptions. These eruptions must have happened relatively recently because, being volatile molecules, they should sublimate and be lost to space on short timescales.

Within the nitrogen ice of Sputnik Planitia there exist some clear signs that heat escapes from Pluto's interior, providing further evidence that the dwarf planet is warm enough to theoretically fuel cryovolcanoes. The features of interest are kilometre-wide, hexagon-shaped 'cells' within the broad, icy nitrogen plains. These are thought to show the effects of convection, a bit like a big icy lava lamp. Each cell is a block of ice warmed by Pluto's internal heat from below, causing some of the ice to upwell and form a mass that is higher in elevation. Yet, on the surface, where the ice is colder and therefore denser, it drifts out and circulates down, making a trough. There is constant overturn of materials in this region. The heart of Pluto appears to be alive.

Intriguingly, similar features are seen on Charon, Pluto's largest moon. Not only is it relatively crater-free, just like Pluto, but it also has landforms such as cliffs and canyons cutting across its middle, stretching for over 900 kilometres (600 miles). It also has ammonia on its surface, as seen by the Gemini Observatory in 2007, indicating that it might still be cryovolcanically active, otherwise, as we saw with Pluto, the molecule would be expected to break down over relatively short timescales. If the ammonia is due to cryovolcanism then scientists suggest it might be produced by episodic explosive eruptions such as plumes, although none have yet been witnessed.

Thanks to all the evidence of cryovolcanic activity observed on Pluto, scientists think it could also house a global liquid ocean below its icy crust, which may have persisted for billions of years. It is still unknown what exactly Pluto's subsurface ocean would be made from, but it is likely to be rich in either water or ammonia, or a mix of the two. An ammonia ocean could be much colder than one made of water, possibly up to 70 degrees colder because of its lower freezing point. Nevertheless, a liquid ocean of whichever composition still throws into question where exactly Pluto finds its heat to keep volatiles warm enough when they would be expected to have frozen solid at the frigid temperatures of the outer Solar System. The problem is that the explanation we've seen so often at other ice worlds is not thought to be possible here. Pluto is not close enough to another large planetary object for it to experience tidal heating. That doesn't leave many options other than Pluto's heat comes from its rocky core, from the decay of radioactive elements. Yet, conventional wisdom says that Pluto is too small and too low in density to have powered itself this way over the course of 4.6 billion years. Its heat would have seeped out long ago if it weren't insulated in some way. Scientists knew they must be missing a piece of the puzzle – something is helping Pluto retain the small amount of heat it's pumped out from its core. The same goes for Pluto's moon Charon, whose smooth surface and tectonic landforms indicate relatively recent activity too, and it is even smaller than Pluto.

This is where a special insulating surface ice might come into play, suggest scientists modelling the interior of Pluto with computer programs. They think a layer of ice, known as clathrate (or gas) hydrate, exists where the base of the icy crust that surrounds Pluto meets the top of the subsurface liquid ocean. The pressure and temperature in this location could provide the perfect conditions for clathrate hydrate to form, as opposed to standard ice, if there is enough gas present. They propose that methane gas is involved, and that it becomes trapped within the cage-like molecular ice structure. A methane-trapping icy layer could act as a blanket insulating Pluto's radioactive heat emanating from its core, while allowing Pluto's surface, which is exposed to space, to be very cold indeed.

The surface of Pluto must remain cold because otherwise it wouldn't be possible for it to form the various tectonic structures and landforms observed, where it is obvious the icy crust varies in thickness from place to place. If Pluto's subsurface ocean is indeed warm enough to be liquid water (no colder than 0°C, 32°F) – which the scientists say is certainly possible within the bounds of their models – then it could even represent an environment that is not completely inhospitable to life. The fact that such an ocean is likely to have been a long-lived feature of Pluto, and possibly Charon too, is an exciting possibility for the potential for the existence of life elsewhere within the Solar System.

While New Horizons didn't capture an ice volcano erupting on either Pluto or Charon, or a plume shooting into space, it revealed evidence that these most icy of ice worlds are active, extending even further the reach of volcanic activity in the Solar System. It seems that we can find evidence for volcanoes, whether made of fire or ice, in every corner of the Solar System.

## Asteroids and dwarfs

Before we finish our visit to the ice volcanoes of the Solar System, let's make a final stop at the smallest of the rocky worlds, the asteroids. We've seen how some asteroids might themselves have been able to fuel a special type of iron-rich volcanic activity on their surfaces long ago. But it might surprise you to learn that they could also have had cryovolcanoes and, perhaps more surprisingly, that such asteroid ice volcanoes could even be active today.

Ceres is the largest object in the asteroid belt, and thanks to its size it is also classified as a dwarf planet, the only one in the inner Solar System. In fact, Ceres is so large that it has managed to form itself into a sphere, unlike most of the smaller asteroids, which can only approximate a rounded shape at best. But Ceres is not a 'classical' asteroid, based on the older textbook definitions anyway. Asteroids have long been expected to be made of rock, having formed within the 'hot' inner Solar System before, or around the same time as, the planets. In the classical definition they are not expected to contain much water or ice because of their proximity to the Sun during their formation and ever since – it should have boiled away. The thing is, despite being found within the asteroid belt, Ceres is not formed completely of rock. It is classed as a C-type, or carbonaceous, asteroid, as are around 75 per cent of known asteroids. C-type asteroids are made of rock as expected, along with clay, but new research shows that they can also be very rich in water. Ceres' crust is thought to be around 60 per cent rock and 40 per cent ice, for example.

The presence of ice and water on Ceres suggests it formed further out in the Solar System than where it is currently found. It potentially began life as a protoplanet near Jupiter, or even beyond, and got diverted into the asteroid belt later on by the gravity of another larger object near which it orbited. Ceres represents a failed planet that ran out of the necessary heat energy to form it.

As we study more asteroids by telescope from Earth, we are learning that it is not just Ceres that doesn't fit the classical idea of what an asteroid 'should' be. Asteroids in the asteroid belt aren't all rocky and many have been found to be icy, particularly those that orbit slightly further out on the colder side of the asteroid belt, towards the orbit of Jupiter. Yet they are still asteroids; the new knowledge just tells us that we need to widen our definition. But you might be wondering how this relates to cryovolcanoes?

NASA's Dawn mission was launched in 2007 to visit two of the largest objects in the asteroid belt: Vesta and Ceres. What it found at Ceres was a surface that, while marked by craters, hosted fewer large craters than expected for such an ancient object. This showed scientists that while Ceres may be a relic from the early Solar System, its surface is far from old. A young planetary surface can only mean one thing: that the object is active within. So, when the Dawn probe spotted a distinctive steep-sided mountain with a crater at the top, now called Ahuna Mons, it didn't take a huge leap of faith for scientists to suggest it was some sort of volcanic dome, but like none seen anywhere else in the Solar System. The problem was that scientists couldn't understand how it got there.

Ahuna Mons is made of ice, making it a cryovolcano, and is thought to have erupted within the last two million years. One of the main lines of evidence in support of recent activity at Ahuna Mons is the bright streaks running down its flanks. On closer inspection, these are found to be made of salt compounds (sodium chloride chemically bound with water and ammonium chloride) and are thought to represent the crystallisation of briny magmas, or cryomagmas, that oozed out of Ahuna Mons just like lava would on Earth. As scientists have continued to pore over the Dawn mission data, they have located dozens more potential cryovolcanic domes on Ceres and predict that one 'pops up' around every 50 million years. The existence of these domes supports the idea that Ceres has been continuously active for a long time, and that its activity is not just a recent phenomenon. They also note that the domes appear to 'relax' back into the surface over time, because the material that composes them, and upon which they sit, is not as solid as rock.

In addition to Ahuna Mons, Ceres hosts some other bright spots on its surface, which were seen by telescope from Earth even before Dawn approached the asteroid belt. Some of these are in a region called the Occator crater, a 145-kilometre (90-mile)-wide crater formed 20 million years ago, and the spots are also thought to be made of the same salt compounds as those at Ahuna Mons. Dawn measured the composition of the salts and found they weren't yet dehydrated (they still contained water in their structure), meaning that they are less than a few hundred years old, as water would be expected to be lost to space in longer timeframes. The revelation is, therefore, that Ceres is an active body which can erupt liquids from its interior.

Scientists didn't really need any more evidence to show them that heat was being transferred from the inside of Ceres to its surface – heat whose source they were yet to understand – but the Dawn probe had more in store. Using high-resolution images sent back from Ceres, scientists mapped out the floor of the Occator crater to reveal a series of unusual hills a few hundred metres in diameter, which didn't look as if they'd been formed by impacts. In fact, they very much resembled the ice-cored hills called pingos that are commonly found in the icy Arctic regions of Earth. Pingos form when water below the surface is frozen and expands, pushing up the ground to form a mound at the surface. The more water supplied to the mound, the larger it gets. And while scientists can't be sure that the hills within the Occator crater are pingos, their morphologies and groupings are close enough that, based on the other evidence for activity at Ceres, they are comfortable linking them with pingos on Earth. The take-home point is that for pingos to form, Ceres must host a reservoir of liquid water below its surface, and liquid water can only be present if Ceres is kept warm in some way.

For the Occator crater, it is thought that the heat required to support the flow of salty cryolavas and a subsurface reservoir of liquid water could have come from the kinetic energy of the impact that produced the crater itself. This event is thought to have occurred around 20 million years ago. Scientists think the impact contacted an already slushy layer just below the surface, causing further melting. The impact is also thought to have created some long-lived cracks in the region that extend deep into the crust and allow brine to continue percolating up to the surface for millions of years subsequently.

The presence of a slushy, or even liquid, subsurface to Ceres even before this relatively large impact suggests that its interior was already being kept warm in some way. The thing is that Ceres, while large for an asteroid, is very small in terms of space bodies, with a mass just one-fifth of the Earth's Moon, so it should have cooled completely by now. Like Pluto, it isn't located in a position next to a giant planet where it can gain heat from gravitational interactions either. Scientists suggest that the most plausible explanation for Ceres' heat is that it is warmed during continued impacts onto its surface, just like the one that formed the Occator crater but on a smaller yet more frequent scale. The slushy layer existing below the surface could then act as an insulator, helping to preserve the latent heat of impact and, in some cases, causing the briny lava to erupt. If this is the case, then there is every chance we might catch more activity on this small world in the future, so it is one to watch. The possibility for a subsurface liquid layer might even present an opportunity for life as we are fairly certain that for biology to take hold, it requires some form of liquid substrate. It is certain that Ceres contains the necessary prebiotic ingredients for life, including carbonaceous materials, and it may even have had the correct environment too.

# CHAPTER TWELVE

# What Next?

The last fifty years has revealed that the space around our precious blue world is full of active and interesting lands. As we've sent increasing numbers of spacecraft out for the first time to explore the planets and moons that share our Solar System, we've uncovered so much more than we expected. We've found landforms that often look just like those on Earth, and revealed unexpectedly young planetary surfaces unmarked by the scars of impact craters. In turn we've discovered that Earth is not the only planetary object with a warm, active interior and this is something that came as a surprise.

We've learnt that even those objects that are barely held within the gravitational grasp of our Sun, existing in the coldest regions of our outer Solar System, can still have geologically active surfaces and even contain the raw ingredients and conditions to support potential life. We've also learnt that size isn't everything, with objects smaller than our cold and dead Moon, and older than 4.5 billion years, still capable of fuelling activity at their surfaces, including cryovolcanoes.

Our knowledge of volcanoes has always been heavily skewed towards those on Earth. Sure enough, our planet has a fantastic variety of volcanic landforms unrivalled anywhere else in the Solar System. Yet by studying our own infinitely more accessible planet, we can use our knowledge about its geological structures to understand the features we are now seeing elsewhere. We can even use Earth as a natural laboratory to help us comprehend the volcanoes we see in space without necessarily having to send spacecraft out to study them in further detail.

It has always amazed me how much we can learn about a place with just a fleeting fly-by. If we consider Pluto, the New Horizons probe sped past, but the advanced instrumentation it carried with it provided scientists with endless reams of data and images to pore over, something that will keep them busy for years to come. The data have revealed a once blurry blob of a world for the first time in high definition, allowing scientists to make inferences about the landforms they see and hypotheses about how they were created.

The next step is for us to go back to these places. For some of these worlds we might be able to send landers, rovers and even rotorcraft, as is already being planned for some of the moons of our Solar System. We mustn't forget, however, how challenging it is to study these alien environments. Space exploration over the last few decades has been full of impressive successes that somewhat obscure the huge amount of effort it takes to make these spacecraft function in such inhospitable and unknown places. Space travel is hard, and to visit the furthest of worlds that exist on the edges of the Solar System presents many challenges. Just getting there can take the best part of a decade, and I know that landing on some of these most far-flung of objects is something I won't see happen in my lifetime.

Nevertheless, I'm excited by what we've found as we've gone out exploring the Solar System. The volcanoes themselves are fascinating, but possibly even better is the fact that they signal that a world is geologically alive. A warm, active planetary interior opens up the very real possibility for life elsewhere in the Solar System. We've seen so often that many of the objects sharing our Sun are formed of plenty of the same ingredients as Earth, containing the basic building blocks to make complex life.

Whether life exists elsewhere in our Solar System or not may never be proven, but I find it comforting to think that we may not be alone. And surely if life doesn't exist here within the environs of our star, then it might do somewhere else in the seemingly infinite expanse of the Universe, where countless other worlds exist around countless other stars. They may host planets or moons that are just like Earth, or others so wonderfully different that we can't even begin to imagine them. For example, the exoplanet 55 Cancri e, it is so hot that scientists suspect it is broadly molten in places, a sure sign of an energy-filled interior. But we should also consider the moons that could be orbiting such exoplanets. While they are harder to spot, because they are so small and distant, we do know there is one orbiting an exoplanet called WASP-49 b that is thought to be erupting. Scientists refer to it as an active 'Io-like' moon, a small, rocky world in a star system far away. With the potential for active worlds in every corner of the Universe, surely there is a high chance of life somewhere else?

Whatever the case, our Solar System is still evolving, albeit at a geologically slow pace. It is not frozen in time, simply preserving a record of the past. Volcanic activity will continue to shape the surfaces of many of the planets and moons surrounding Earth in the coming years, extending into the billions of years, until our Sun dies and humans have long since departed. On our relatively short time on this planet we are privileged to have been able to harness our intelligence and build the tools to open our horizons and explore beyond our blue planet, and I can't wait to see what else we will discover in the future.

# Acknowledgements

As many an author has said, writing a book is a rather solitary endeavour. Yet this book has kept me company, moving countries with me and my family from California, USA, to the UK, and putting up with three house moves. It got put on the back burner for a year while I flew off to write a planetarium space show in New York, but was there waiting for me when I returned. It has even dealt with a global pandemic. Nevertheless, this book eventually got written and the breaks from writing I've had to take over the two and a half years working on it helped to give me some perspective.

I absolutely love the process of writing, shutting myself away in the office for hours on end to research and write about these fascinating Solar System worlds. Towards the end of this book, when the Covid-19 lockdown began and I had my daughter at home all day, it would have been easier to take another break from writing. But, thanks to the commitment of my amazing husband, who took the reins on childcare at the weekends, I was able to continue pushing through the words to complete my first draft. This gave me such joy (and a break from the endless hours of entertaining my daughter) and a big boost to know I had the space to think during such unprecedented times. This book kept me going, as well as Joe Wicks' PE lessons on TV!

In terms of proofreading, first thanks will always go to my father Colin Starkey, who is, as always, the first to have read the first draft at the stage when I'm still not really ready for anyone to see any of it! Then I've had a whole raft of friends who have kindly offered their time to check the English and, just as importantly, the science. Whether they read one chapter or the whole thing, I am truly grateful for their help to make this book better. So, thank you to: Tess Mize, Wendy Tomlins, Rhian Meara, Ashley King, Jack Wright, Helen Cooke and last, but not least, Susanne Schwenzer.

I thank my daughter for being such a ray of light during the hard times. Every time she excitedly recites the names of the planets in order, she reminds me why I do this. I want everyone to love learning about space as much as I do, and I hope that my readers will enjoy this book and feel inspired to learn more.

# Index